Electronic Protection and
Security Systems

Electronic Protection and Security Systems

Second edition

Gerard Honey

Newnes

OXFORD   BOSTON   JOHANNESBURG   MELBOURNE   NEW DELHI   SINGAPORE

Newnes
An imprint of Butterworth-Heinemann
Linacre House, Jordan Hill, Oxford OX2 8DP
225 Wildwood Avenue, Woburn, MA 01801-2041
A division of Reed Educational and Professional Publishing Ltd

℞ A member of the Reed Elsevier plc group

First published 1996
Second edition 1998

© Gerard Honey 1996, 1998

British Library Cataloguing in Publication Data
A catalogue record for this book is available from the British Library

ISBN 0 7506 4229 7

Library of Congress Cataloguing in Publication Data
A catalogue record for this book is available from the Library of Congress

Typeset by Avocet Typeset, Brill, Aylesbury, Bucks
Printed and bound in Great Britain by
Biddles Ltd, Guildford and King's Lynn.

PLANT A
TREE

British Trust for
Conservation Volunteers

FOR EVERY TITLE THAT WE PUBLISH, BUTTERWORTH-HEINEMANN
WILL PAY FOR BTCV TO PLANT AND CARE FOR A TREE.

Contents

Preface

Electronic security is a basic need since it is impossible to live a full and rewarding life without a feeling of well being for ourselves, family and possessions. Certainly by having effective protection it helps to buy the one thing that cannot normally be purchased – 'peace of mind'. Such electronic protection exists in many forms but meets a common goal, that of protecting life, possessions and premises.

All the means described in this book are based on an electronic principle of operation and to some extent are governed by National Standards and Codes of Practice for numerous features. These requirements are therefore shown to help understand how they influence the final planning, installation, maintenance and records and also system and device functions.

This book shows the various types of electronic protection and security systems that are generally available and the factors that influence the use of them. The basic concepts are explained so that the individual can select the appropriate means to suit specific requirements.

Gerard Honey

Acknowledgements

We would like to thank the following firms for their invaluable help
in supplying product information for this book:

BPT Security Systems (UK) Ltd
The British Security Industry Association (BSIA)
C&K Systems Ltd
Fermax Electronica S.A.E.
Group 4 Total Security Ltd
Maplins Electronics PLC
Menvier (Electronic Engineers) Ltd   Menvier Security Ltd
RS Components Ltd

1 Intruder alarms

The role of security and protection is a serious one. The existence of crime makes us all potential victims and obliged to recognize that to some extent we are all vulnerable.

Crime is all too graphically described every day in our newspapers and on our televisions. There are now so many attractive targets for the burglar – TVs, hi-fis, videos and computers are examples of modern items, along with more traditional objects such as cash, jewellery and antiques.

As a result more of us must consider ways of protecting our family and possessions. This can be done in a number of different ways but installing a security system is probably the most effective method.

Everyone who owns property or goods can be a victim of the thief and no one is immune from the possibility of being violated by the burglar. Therefore it is up to the individual to reduce the chances of becoming a statistic. It is a fact that a great deal of break-ins do occur during daylight hours even if the house may be empty for only a short while. The thief can always make the most of a quick opportunity so it is essential to make life difficult for the intruder and prevent him/her from being able to act in an expedient manner.

Very often an intruder alarm system can be integrated with a fire alarm system, but in this chapter we intend only to show how property loss can be prevented by means of an intruder or burglar alarm system. A basic intruder alarm system consists of a number of detection devices, a control panel and signalling devices such as sirens or bells. The system may also be connected to a telephone line to provide remote signalling. Every element in the system has its own specific function and is carefully designed to blend into its surroundings to give a consistent, reliable and trouble-free performance.

1.1 Detection devices

Detection devices are intended to recognize attempted intrusion. They consist of components which are strategically placed around the premises to be protected, dividing it into zones or areas. For instance in a house, the living room, kitchen, garage, main bedroom and secondary bedrooms can all be planned to operate on different zones.

Specific zones can be protected or left unarmed at any given time, depending on the configuration the homeowner chooses. The method of establishing zones is dealt with at a later stage, but at this point we shall look at detection devices in some detail.

Detection devices play many diverse roles but fall into two main categories.

- *Perimeter detectors.* Such devices sense attempted intrusion.
- *Motion detectors.* These devices sense movement in a protected zone or area.

Magnetic reed switches

One of the most basic perimeter devices is the magnetic reed switch. Switches intended to protect opening doors or windows are normally of the closed magnetic variety. Basic versions are encased in plastic and consist of a metallic arm which closes the contacts when a magnet is held in close proximity to the switch face. For heavier duty, metal cased switches are also available for use in harsh atmospheres or where the switches may be subject to abuse, such as in an exposed position. When the magnet is moved away from the switch, such as when the door or window is opened, the moving contact, being no longer attracted, leaves the closed position, producing an open circuit in the switch mechanism or alarm contacts. It is this open circuit that is recognized by the control panel which then triggers the signalling devices to produce the required alarm.

An inert gas surrounding the mechanism keeps the switch contacts in good condition and resists corrosion. The magnetic type sensor or reed switch is inexpensive, reliable and has a rated life of 10 million operations, which is far in excess of the operating life of more 'hi-tech' items used in home security. The magnetic reed switch normally features a terminal block of five terminals unless a fully sealed version is required. In the latter case the cables are pre-wired within the unit which is then welded together as an assembly to make it tamper proof.

Magnetic switches are either surface mounted or flush mounted. The choice depends on the opening that is to be protected. Windows generally use surface mounted types whereas flush mounted units are preferable for doors and particularly for wooden doors. In the case of flush units, the switch is mounted within the door frame and the operating magnet set into the door.

All reed switches have a specified opening gap, i.e. distance between the switch and operating magnet. This should be checked before installation using generally available test equipment – an ordinary multitester will suffice – and then verified after installation when the system has been fully commissioned.

Flush unit switches when used on internal and external doors are fully hidden, making for a neat installation which also has a high security effect. Areas of the premises requiring protection are isolated by the system, since doors cannot be opened with the alarm system switched on. If the wiring to the devices can be effectively concealed an extremely tidy installation can be achieved.

Typical domestic intruder alarms generally feature essential external doors and a number of ground floor doors protected by reed magnetic switches. If any external doors are subject to particularly damp or difficult weather conditions then surface switches are advocated. These switches are not concealed. They are often found protecting windows, but are also used on patio type doors or garage doors. The switch unit is mounted on the stationary frame and the magnet on the movable part such as the moving window. The movable part must stay in its closed position or the desired open circuit will occur in the switch which will be recognized by the control panel.

When surface mounted switches are to be used, care must be exercised since these devices remain visible. However, their effectiveness is unquestioned providing that they can be made inaccessible under normal circumstances, and when used in conjunction with flush switches elsewhere make up the nucleus of a good base alarm.

Reed switches make use of their five terminals to join and run wires in series with one another. Certain terminals are used for the alarm contacts and others are used to provide a continuous tamper loop, which can be taken back to the control panel for constant monitoring of the system. The terminals shown in Figure 1.1 for the alarm loop can be recognized on the switches by the reed legs which are connected to them by the manufacturer.

The other widely used reed switch is the roller shutter contact which is fully sealed within either a robust metallic or plastic enclosure and comes pre-wired. It is used in heavy duty applications where the door can be subject to slight movement even when locked, also its powerful magnet helps cater for any slight door misalignment.

As the name suggests these switches are used on door and window roller shutters generally with the switch floor mounted close to the vertical door rail channel. They are also often found on large wooden barn doors – the flying leads terminate in a tamper protected junction box which is connected to the alarm wiring which runs back to the control panel.

There is a huge variety of reed switches available for use in the

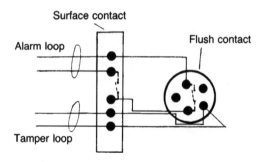

Figure 1.1   *Alarm loop*

intruder alarm industry yet their application and installation is a very straightforward process – one should almost never see a burglar alarm system that does not use a reed switch somewhere within its confines.

Another important device which tends to use reeds and a similar wiring method is the personal attack button (PA). This is discussed in Section 1.5.

Originally the magnetic switch was found allied with the pressure sensitive mat in many intruder systems. The pressure mat, which is a normally open contact device, was placed under carpets and responded to pressure being put upon it, for example by an intruder. In practice it is rarely used now but it is important to understand how a normally open device is configured and wired back to the control panel. Figure 1.2(a) shows the wiring of a normally open detector whilst the significance of the alarm and tamper loop at the control panel is shown in Figure 1.7.

Having an understanding of the wiring of the normally closed and normally open detectors can then enable the reader to explore the other more advanced forms of wiring the detection devices using resistors. Figure 1.2(b) shows a double pole with end of line resistor (EOL), Figure 1.2(c) shows dual operation and Figure 1.2(d) depicts a dual purpose method with variant value resistors.

We can say that Figure 1.1 shows the traditional standard wiring method using a positive alarm loop with the switches or detectors wired in series, with the tamper loop as a separate pair, wired as a negative loop running in the same four core cable. Figure 1.2(a) is for normally open detectors with the tampers also to be in series but with the alarm contacts in parallel using four core cable. At the control panel the alarm loop is actually wired across the positive and negative loops as shown in Figure 1.7. The methods employed in Figures

1.2(b), (c) and (d) are for use with particular control panels which may employ a different value of resistor for monitoring purposes. These are all widely practised wiring methods and should be understood.

Figure 1.2(b) is not unlike Figure 1.1 but has an end of line resistor (EOL) at the end of the positive alarm loop. The control panel can then analyse the impedance and determine if the circuit is open and in alarm, if it is normal with the full impedance, or if the loop has been shorted and the impedance is no longer capable of being recognized. It is not complex to those familiar with the standard wiring methods as it uses a four core cable with separate alarm and tamper loops. The only disadvantage lies in that four core cable must be used, although two core cable can achieve the same effect by using one of two methods of dual operation/purpose.

In Figure 1.2(c) all devices are wired as normal but with a 4k7 resistor being fitted in series. The tamper is also wired as normal but with the same value resistor being fitted in series. By this means two core cable can be used so that supervision is achieved with cable reduction. The zone loop as depicted is wired back at the control panel to designated terminals for that purpose.

Figure 1.2(d) is a dual purpose loop with an end of line resistor that can be used with both normally closed and normally open detectors. In this case the end of line terminating resistor is 2k2 in value and at the end of the loop. Higher value resistors are employed across each pair of series contacts. A continuous signal will continue to pass

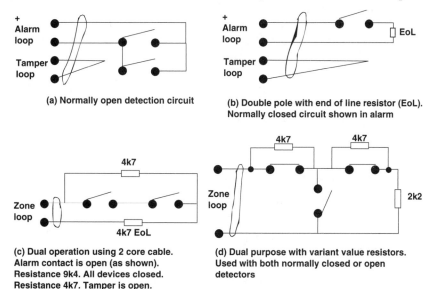

(a) Normally open detection circuit

(b) Double pole with end of line resistor (EoL).
Normally closed circuit shown in alarm

(c) Dual operation using 2 core cable.
Alarm contact is open (as shown).
Resistance 9k4. All devices closed.
Resistance 4k7. Tamper is open.
Loop is open circuit

(d) Dual purpose with variant value resistors.
Used with both normally closed or open detectors

Figure 1.2 *Detector circuits*

through the circuit irrespective of the operation of the detection devices although it will vary as the detectors are tripped. It can be seen that the circuit is also monitored for tamper open circuit disconnection so it follows that the circuit should not go open.

When the control panel is set, if any detector is tripped the loop resistance will be affected and the current changed, causing an alarm to be generated by the control panel. The system will equally respond to a short circuit which is effectively similar to the normally open detector going closed.

For those systems we have described using resistors the wiring back at the control panel will be specific to the equipment being used and so will differ from the basic circuit in Figure 1.7. Many control panels can indeed be programmed to allow the installer options regarding the use of more traditional wiring type such as those shown in Figures 1.1 and 1.7 or to move to detection circuits with resistors. We must add that when resistors are used these must be placed within the detector itself.

Window foil

In cases where an intruder may attempt to leave a window in its closed position or for non-opening windows, the installation of foil to protect the glass from being broken may be considered. Window foil is used in applications where its rather untidy appearance is not important. In practice window foil is extremely effective and works on the principle that if it is broken anywhere along its length or if the termination is knocked off it will produce the desired open circuit.

Window foil is available in different grades and sizes and has a self-adhesive back to allow ease of fitting. It is terminated using unique cabling blocks from which the main wiring is taken back to the control panel. Foil is an inexpensive method of protecting areas but can be replaced by more up-to-date vibration or break-glass detectors where aesthetic value is important.

Vibration/impact detection devices

For physical structures which may be subject to attack in the event of forced entry, solid state vibration/impact detection devices may be employed. These are designed for mounting on doors, windows, walls, roofs or safes and may be installed in any orientation.

Vibration/impact detection devices use a piezo-electric crystal sensor which produces electrical impulses when subjected to vibration or sudden impact. The impulses are formed into a signal and the unit features an integral analyser and sensitivity control.

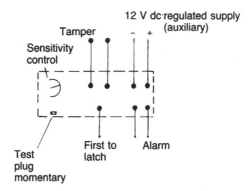

Figure 1.3 *Basic terminal board*

In an intruder alarm system the unit is wired to a 12 V dc regulated supply provided by the control panel. The alarm output of the detector is derived from the isolated contacts of a normally energized relay that also detects any supply failure to the device.

A useful application of the impact detector is on doors through which an intruder could attempt to gain entry by forcing out a panel or forcing the door from its hinged side leaving any protecting magnetic reed switches *in situ*. The impact detector, when complementing reed switches, prevents any attempt to enter by such means.

With vibration/impact detection devices it is important that as much of the protected surface is in contact with the base plate as is possible with minimum distortion. Tapping the protected surface causes the device to go into an alarm mode and this is shown by a light emitting diode (LED) response which can be noted. An integral sensitivity control can then be adjusted so that the device operates at the desired range of impact response, taking into account environmental conditions.

In view of the fact that the device, once alarmed, will reset and restore the security system to its original mode after a prescribed period, it is important to know which detection device has caused the alarm condition if a number of detectors are used. This is done by using devices that have a 'first to latch' option feature which maintains the LED in its illuminated condition on the unit that caused the alarm condition.

Impact detectors are available in many different styles. A basic terminal board is shown in Figure 1.3 and is typical; variant types do adopt a common wiring method and application use. The assembly shown is held within a tamper protected enclosure.

Reference should be made to Figure 1.7 for wiring at the control panel. A momentary test plug featured on most of these detection devices can be used in conjunction with the sensitivity control to set the level of vibration or impact desired for the unit's operation.

Break-glass detectors

Break-glass detectors are suitable for fixing on standard window glass or plate glass; they detect and respond to a frequency in the region of 150 kHz when the glass is broken. They are secured in position by adhesive and, being passive (i.e. they do not generate external energy), any number of detectors may be installed on the same window pane without any risk of interaction.

Like vibration detectors, break-glass detectors use a fail safe, normally closed construction. This output is often connected into an anti-tamper 24-hour guard loop provided by the control panel. This loop is permanently active so if the detector goes into alarm even with the control panel in the off condition, an audible response will be made by the panel. A latching LED indicator can once again identify the detector which has triggered the alarm condition.

Other forms of break-glass detector which respond to the ultra sound created by breaking glass are available and can be installed on the window frame, on the ceiling or even up to 2.5 m from the window being protected. Such detectors are most suitable for display windows as they do not need to be secured to the glass. These are often supplied both with normally closed alarm and tamper outputs plus latching LEDs where multiple devices are employed.

The break-glass detector is powered by a 12 V dc regulated supply from the control panel in the same way as the vibration/impact detector. All devices powered in such a manner are generally called auxiliary devices and all the detection units which follow in this section fall into that category. The 12 V power terminal to which they are connected is often referred to as the auxiliary (aux). This is shown in Figure 1.7.

Passive infra red (PIR)

As stated earlier, detection devices are normally classed as perimeter detectors or motion detectors. The most common motion detector in use that senses movement in a protected area is the Passive Infra Red (PIR).

Modern PIRs can blend into any domestic environment. They offer exceptional detection capabilities and false alarm immunity if correctly selected and sited. A vast array of types are currently available. PIRs offer great potential, are not expensive and can form the nucleus of an up-to-date highly effective intruder alarm system.

In the first instance one should recognize the significance of the sensor which uses a principle of passive infra red. The PIR sensor detects a rapid change in radiation within its area of coverage. The passive infra red sensor measures the infra red energy radiated from every object within its field of view and monitors changes in the level of radiation. If this change is large enough and rapid enough the sensor will respond. Separate elements are generally employed so that when there is a difference between infra red energy detected by each of these elements the sensor will trigger. Such balanced detection offers excellent immunity to false triggering that could be caused by quick changes in background temperature whilst not compromising the detection of a real intruder. Pulse count triggering provides a further improvement towards false alarms. When the circuit first senses an alarm (pulse) it goes into alarm standby for a prescribed time (often 20–30 seconds). If while in this mode a second alarm event occurs (pulse 2) the detector will immediately alarm; if there is no second pulse then the detector will return to its normal state. Additional pulses can also be selected in areas subject to pulsating heat sources. The most common type of sensor forms a volumetric coverage angle in the order of 90° using finger patterns. Various distances can be attained over which the PIR is effective but curtain or corridor detection patterns which need an effective minimum length of some 20 m tend to employ a much reduced angle form. Section 1.6, 'False alarms', gives mention of variant PIR patterns intended for use in areas where domestic pets may be present.

To fill a space and ensure any intruder is detected by the volumetric PIR sensor it is best placed at ceiling height, not in direct sunlight and in a corner position. PIRs are more sensitive to movement across the detection zones rather than movement towards the sensor and placing them in a corner eliminates the blind spots that can occur around the sides of the device. Figure 1.4 illustrates the most popular coverage patterns. If reference is made to Figures 1.9, 1.10 and 1.11 one can see the effectiveness when corner mounting is employed. Equally if one is concerned about a possible blind spot under the detector then a variant form employing a creep zone, to eliminate that position, can be used.

PIRs must be carefully positioned if false alarms are to be avoided and common sources are detailed in Section 1.6. All PIRs require a regulated 12 V supply from the control panel to provide power to the device. Most devices operate in a normally closed state which goes open circuit when the unit is tripped by an intruder. This response is

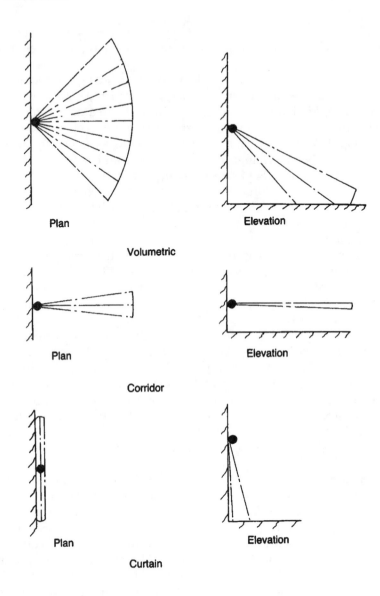

Figure 1.4   *Popular coverage patterns*

recognized by the control panel and an alarm is generated. The typical wiring of these motion detectors is shown in Figure 1.7.

PIRs feature a walk test LED so that the installer can check the operation of the unit, once the wiring has been competed, by walking in the protected area. The LED will operate at each activation. Many PIRs have a facility to allow the LED to be blanked once the coverage pattern has been proved. This stops a potential intruder attempting to gather information into the detection coverage of a particular device. It is blanked by either fitting a special cap over the LED or removing a designated jumper on the printed circuit board. PIRs are fully tamper protected to stop any unauthorized removal of the cover or lens, therefore preventing any sabotage.

The choice of PIR is extremely wide and because they form the nucleus of intruder alarm detection, being the most extensively used sensor, it would be useful to summarize their patterns and the optics they use. These patterns are essentially volumetric, corridor, curtain and pet alley.

In simple terms volumetric means a standard fan-shaped zone generally filling a space with a detection angle in the order of 90°.

Corridor means fewer zones across a narrow width but having a longer detection capability. This can be as long as 40 m.

Curtain defines a vertical pattern running parallel with the area to be protected, creating a solid protected zone up to some 20 m.

Pet alley is self-explanatory and is that pattern in which the floor area is left uncovered. Its range is often similar to curtain.

In some detectors the pattern can be changed by slotting in an alternative lens. These are usually fresnel optic lenses which are plastic with a surface of moulded ridges. Some manufacturers do not favour this approach however and prefer to supply a range of different detectors each with its own permanent lens because they feel that otherwise focusing accuracy is diminished. The alternative to the use of fresnel lenses is to adopt mirror techniques which provide a sharper focus although these optics are, in the main, more expensive. Although mirror systems cannot be changed, coverage can be modified by the application of masks. The choice of lens or application of masks to avoid certain areas being viewed is important. Troublesome heat sources fall into this category and are discussed later in the section on false alarms.

In areas such as those found in the petrochemical, pharmaceutical, explosives and paint industries special PIR movement detectors are used. These meet the demands of BASEEFA (British Approvals Service for Electrical Equipment in Flammable Atmospheres). This is achieved by using the devices in conjunction with zener barriers, which restrict voltage and current to a level where there is insufficient energy in the hazardous area to create an electrical spark that could ignite the surrounding atmosphere. The barriers are installed outside of the hazardous area in the PIR control system and allow only very low energy signals to pass to and from the detectors.

Ultrasonic detectors

Another widely used volumetric detection or movement detection device is the ultrasonic detector, although currently these are not quite as popular as PIR units. The ultrasonic detector operates by filling an area with a continuous high frequency whistle just above that of sound. This cannot be heard by the human ear. The whistling effect bounces from almost all surfaces and providing the frequency of the tone received is the same as that transmitted it can be assumed that all objects are in a static condition. However if any movement occurs in the protected area a part of the signal will experience a change in frequency. This is called the doppler effect. A shift frequency over a certain duration of time will trigger the alarm. In order to ensure that false alarms cannot be caused by turbulence in the air or the movement of curtains or the like, measures are taken in the detector design. Advanced ultrasonic units are provided with an anti-masking feature which prevents any obstruction being placed before the unit. In these cases it will signal an alarm condition if anything is too close. This also ensures that a mask cannot be placed over the face of the detector.

Unfortunately some installers attempt to standardize on their use of detector rather than viewing each application independently. There are many instances where the ultrasonic principle is the best option.

The ultrasonic detector does not have its sensitivity impaired by the ambient temperature and it detects an approaching body more effectively than a PIR. Also it readily neglects pulsating heat sources which can prove troublesome for an infra red detector.

One disadvantage of the ultrasonic detector is its physical size. It is generally larger than a competing PIR and it has a limited range which usually only extends to some 7–8 m.

As with PIRs a single LED is provided to show any background disturbance, walk test and alarm memory indication. The walk test allows the installer to determine the coverage pattern and to avoid any false alarm hazards.

The typical coverage in free space for a general domestic/ commercial ultrasonic unit mounted at a height of approximately 2.5 m is shown in Figure 1.5.

The wiring method of the ultrasonic detector (Figure 1.7) to the control panel is similar to that of the PIR device, indeed all of the following detectors – combined PIR/ultrasonic, microwave, dual technology, combined acoustic PIR, active infra red, capacitance and ceiling mount PIR detectors – follow the exact same practice of alarm connection. Although more specialized in use the ultrasonic detector does have a wide application and hence we should understand its role and importance in the security industry.

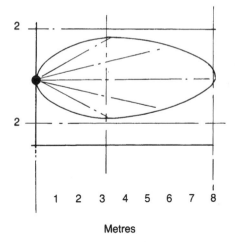

Metres

Figure 1.5   *Plan view*

Combined PIR/ultrasonic detectors

Many PIR devices employ an advanced double edge signal process and dual element pyroelectric detector for reliability and high false alarm immunity plus different levels of sensitivity. Exceptionally low false alarm probability is ensured since an intruder must cross each half of a detection zone, creating signals on both halves of the dual sensor, before an alarm is generated. This eliminates false alarms that can be caused by temperature changes in static objects such as boilers.

In certain cases where environments could still be troublesome to PIR detectors, however, combined PIR and ultrasonic principles are used. These units are suitable for most space detector applications and use the combined techniques which react to different forms of stimuli. Hence an alarm condition can only be made in response to a stimulus to which both detectors will react. The alarm signals from each detector are combined by a logic circuit in a way that they must alarm within a preset time period before the alarm signal is given.

A typical device will feature the passive infra red detector behind a lower front window to provide the normal volumetric protection. The ultrasonic portion of the unit will be sited in the upper part of the detector. Careful design of the unit permits the installer to optimize the mounting angle with tamper protection preventing any attempt to remove the front cover to get access to the unit or indeed to remove the unit from its mounting. Walk test indicator LEDs allow separate test indication for the combined unit and for the independent PIR and ultrasonic elements.

Microwave detectors

The envelope pattern of microwave detectors is very similar to that of the ultrasonic device but the coverage of the microwave unit is much greater and can extend up to 30 m. Microwave energy is transmitted into a broad pattern throughout the protected area at an extremely high frequency range. Using the Doppler effect any frequency change of the signal reflected is analysed to determine the size of the object and the distance moved. An alarm output will occur when a large object, such as an intruder, has moved a predetermined distance. As with ultrasonic units an anti-masking facility is generally available.

The microwave principle is often adopted as a security aid for outdoor applications and, although its nominal range is 30 m for an average size human being moving towards or away from the detector housing, a well-designed unit can detect a large vehicle at a distance of 100 m. Basically, the range is governed by the target size.

Outdoor microwave intruder detectors are particularly suitable for use in conjunction with closed circuit television (CCTV), security lighting and other switched devices to provide a reliable outdoor trap protection system for compounds, yards and other similar sites. Indeed microwave detectors do have significant advantages, being unaffected by air currents or noises that can prove troublesome to its ultrasonic counterpart.

Dual technology detectors

An exceptionally low false alarm rate can be achieved by using integrated infra red (IR) and microwave elements. This type of device is often termed dual technology although the term is sometimes also applied to combined PIR and ultrasonic units.

An excellent detection performance is assured by the combined microwave and infra red unit making this form of detector ideally suited for use in remote sites where false alarm call outs would be expensive and troublesome. High risk areas, potentially troublesome environments and sites requiring coverage over large areas are well served by such units.

Coverage is obtained by using a wide-angle microwave pattern matched to well-recognized infra red zoning. Coverage of large areas of up to 20 m and 15 m is typical and the use of dual technology detectors means that fewer sensors are needed, consequently less wiring is necessary, installation and maintenance are easier and there is lower power consumption.

The dual technology detector is often found giving space protection to large open buildings, barns and warehouses with the external door security provided by roller shutter contacts.

Combined acoustic PIR detectors

For exceptionally good detection and freedom from unwanted alarms intelligent combined acoustic and passive infra red sensors may be considered. These sensors, which are in one enclosure, combine two established intrusion detection techniques with individually adjustable sensors to suit both the environment and level of required security.

The acoustic sensor used to detect breaking glass has an independent LED to that of the PIR sensor and is fully adjustable so that it may protect windows a short distance away from the detector. The PIR covers a much larger area, usually in the region of 10–12 m. Detector modes allow independent operation as each sensor has an individual alarm relay output.

Alternatively, for confirmed detection, the digital logic in the detector can monitor both elements and only generate an alarm signal if acoustic detection is followed by PIR detection. This would apply where the requirement is for the break-glass condition to be verified as entry by the PIR detection. These combined type detectors can thus satisfy both perimeter and space protection needs by the installation of one unit only.

Active infra red detectors

When large glass surfaces, walls and ceilings or showrooms and shelves, etc. need protecting, active infra red barrier detectors may be used for both indoor and outdoor applications. These are not affected by direct sunlight and have a range of 20–30 m depending on where they are used.

Such devices use a transmitter and receiver principle to project an invisible infra red beam across an open path. These systems are reliable and have a long service life. Further operating security is achieved when the receiver is exposed exclusively to the beam of one transmitter. The capacity of transmission is not split and secure operation is guaranteed. Detectors can normally be mounted on top of each other so that the effective distance between the beams can be reduced to 10 cm apart. An intruder crossing between the transmitter and receiver destroys the invisible beam which causes an alarm signal to be generated at the control panel.

Capacitance detectors

Capacitance detectors work on the change of electrical capacitance between a protected object and earth. Normally this is produced by the hand or body capacitance of an intruder, bearing in mind that the human body possesses both electrical resistance and capacitance.

Capacitance detector units tend to be used to protect metal safes, cabinets containing important files or valuables, or other metallic objects. Items that are not themselves of metallic construction can be protected by fixing metal foil strips to them.

Capacitance detectors are sensitive to either an increase or a decrease in capacitance so any removal of the object or foil can be sensed. Slow changes in capacitance are compensated for and so are problems of electrical interference. Setting up the unit is achieved by using the integral LEDs, with sensitivity being adjusted in steps to suit the pertaining environment. Tampering with the sense lead between the object being protected and the detector will also generate an alarm.

Ceiling mounted PIR detectors

The reader will now appreciate the vast array of intruder detection devices available and should be in a position to select a basic device to suit any given area. However, it will be found in practice that many variants exist for each basic detection device. For this reason, ceiling mounted products are becoming more popular especially the ceiling mounted PIR which has distinct advantages in many applications.

Ceiling mounted PIRs, although slightly larger in size than wall mounted devices, do have certain notable factors in their favour. Direct sunlight falling onto the sensor is avoided and the area of coverage can be readily determined. Ceiling mounted PIRs, when used in the office or commercial sector, can blend very easily into the environment and indeed look similar to a light fitting. Mounting on the ceiling ensures that no stacking space is made redundant, which occurs with wall mounted PIRs since obstructions must not be placed in their field of view. Small shops, for instance, often need to stack goods against all walls up to ceiling level and the positioning of standard PIRs can be a nuisance. Ceiling mounted PIRs are also very useful in open plan office sites.

A typical ceiling mounted PIR will consist of a base, containing the terminal block, and the sensor head, which can plug into the base and be secured by screw fittings. The sensor head should be sealed against the ingress of dust and to preclude insects. The detection field through 360° will cover a circle of some 10 m radius with the unit mounted on ceilings between 2.5 and 5 m high. Double activation or pulse count triggering is generally included, whereby an initial intrusion will put the unit on standby for a prescribed number of seconds and an alarm will only be generated following subsequent movement.

High or low sensitivity selection should also exist – low setting is used where potential infra red hazards exist or for small area protection. Other ceiling mounted PIRs do not of necessity have a 360° coverage but have pattern choices of different angular spreads and detection ranges.

Unwanted signals caused by light, rapid changes in temperature and radio frequency interference are excluded by using two pyro elements that must be affected by alternating signals generated by an intruder rather than by simultaneous signals.

The reader will now have come to appreciate basic intruder detection devices. When the control panel recognizes a change in state of the detection device it will generate an alarm by providing an output to the signalling equipment. Perhaps at this stage we should look at the simple electric circuit as shown in Figure 1.6. We can then go on to look at the electronic detection circuit.

The simple circuit illustrated in Figure 1.6 consists of a power source, wire connections to complete the circuit, a switch and a signalling device to sound the alarm. In this case the power could be mains ac (alternating current) or self-contained dc (direct current) derived from a battery. However, power will be provided in most instances by a combination of dc taken from a mains transformer plus battery support in the event of mains failure. It can be seen that by operating the normally open switch the signalling device will be triggered and the alarm sounded. Such a circuit can only conduct current, dissipate power and sound the alarm when the switch is closed. A simple system such as this is often used for energizing lights or ringing bells when windows or doors are opened. The orientation of the switch is set to cater for such operation.

In the standard form of an alarm system an electronic latch is required once a signal has been applied. This is generally provided by the control panel. The base electronic detection alarm circuit is essentially as shown by the wiring diagram in Figure 1.7.

The alarm loop is made up of normally closed and normally open detection devices. When maintained in the original mode, with the control panel in the 'off' position, no trigger will be sent to the sounders or signalling equipment by the control panel. The tamper

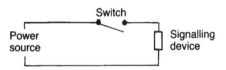

Figure 1.6 *Simple electric circuit*

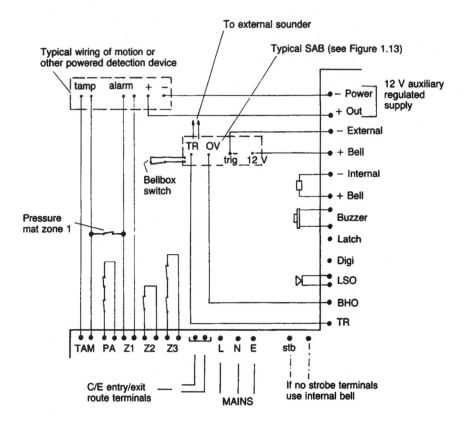

Figure 1.7 *Base electronic detection alarm circuit*

loop which runs in the same cable harness as the alarm loop is continuously monitored. This is known as a 24-hour circuit or tamper loop and ensures that no disconnection can be made to any of the wiring even with the control panel in the 'off' position. The alarm loop is only active when the control panel is in a set or 'on' condition. Normally open devices, such as pressure mats, are connected across the alarm and tamper loops so that any short circuit across these points will cause the control panel to trigger the signalling device when it is in the set position. The control panel will then latch electronically to maintain an alarm condition even though the detection device may only have given a momentary pulse.

The tamper loop is also important because it is negatively biased, and since the alarm loop is positive, it is not possible for an intruder to short out the wiring of reed switches in order to close their circuits and defeat them, because in presenting a short across the different polarity loops an alarm is generated. Certain detection devices are however destroyed once activated and this is extremely troublesome for both the owner of the premises and the installer. Clearly detectors that give only a momentary pulse and do not need attention or to be reset are superior. At an earlier point we looked at window foil which is destroyed when the window is broken. In keeping with foil some burglar alarm systems do still employ window tubes. Although long established the window tube is still acceptable and, when installed, looks like steel bar physical protection. In practice it comprises a series of vertical tubes spaced at approximately every 100 mm; threaded through them are single strands of brittle wire, anchored at the ends of each tube. The alarm is activated when the tubes are forced apart by intrusion, breaking the brittle wire in the tube and opening the circuit to which it is connected. This idea is somewhat similar to door wiring which consists of a pattern of brittle wires irregularly stapled to the inside of a door at risk. The wire is then concealed and protected by fitting a sheet of plywood or equivalent over the wiring on the door. Cutting a man-size hole in the door will destroy at least one of the sensing wire loops.

Clearly detection devices such as foil and tube or door wiring go open circuit in alarm, indeed the majority of detectors do operate in this mode. However the intruder alarm control panel does accept devices which switch to closed circuit in alarm – for example the pressure mat. The panel must therefore cater for different switching modes and be capable of isolating detection devices that are destroyed or ineffective. Equally it must be able to select or deselect detectors at different times and to automatically stop any signalling, audible, visual or remote after a selected preset time.

The heart of the intruder alarm system must therefore be the control panel. This monitors the state of the alarm and anti-tamper circuits and generates the alarm condition when in the set position and following a detection device changing its state. The cabling running between the control panel through to the detection devices and from that point to the signalling units is either 4 core – for devices such as switches or pressure mats that do not require a 12 V regulated supply – or 6 core cable – for devices that do need connection to this supply. General cable contains flexible wires each having some seven strands of 0.2 mm tinned annealed copper insulated wire conforming to BS 6360. The insulation sheath is PVC and the nominal overall diameter for 4 core or 6 core cable is 3.5 mm or 4.1 mm respectively. Other multicore cables are also available but their use is somewhat more specialized. It is not essential that these cables are totally concealed or under floors, etc. but they are often run under carpets close to skirting boards as this can save disruption or damage to decor.

1.2 Control panels

The growth of the intruder alarm industry has resulted in a proliferation of control panels becoming available and as the selection of panel very much depends on the duty that is required of it, it would be a mistake to generalize.

The panel selected, however, should possess the number of zones that the premises have been sectioned into plus other features necessary to get maximum effectiveness from the zoning. A level of guidance is provided in Section 1.3. The signalling or bell output capacity of the panel is of major importance. The term 'bell' is actually time-honoured usage and is normally used in the identification of terminals used for signalling, but in practice these terminals can equally apply to any form of sounder be it electro-mechanical or electronic.

In certain instances a remote signalling output to a central station over the telephone network is a requirement, using automatic dialling equipment or digital communicators. A short discussion of this duty will be found in Section 5.5. However, it will be found that the generally available competing control panel will have a 240 V ac mains connection and provide a 12 V dc working output by means of the integral transformer and electronics circuitry. Room inside the panel will also be provided for a standby rechargeable battery in the event of a mains failure occurring. The detection devices are wired into the appropriate zone terminals provided by the control panel, but in addition to the number of alarm zones featured, terminals for the entry/exit circuit route must also be provided. This is the route used by the authorized occupants to leave and re-enter the premises once the control panel has been set. It provides a measured time delay, and detection devices that will be tripped during this period must be wired on this circuit. Movement detectors and other energized devices are powered from the 12 V dc regulated auxiliary supply. In the event of mains failure the battery will provide this supply. Batteries are sealed, being essentially maintenance free, and are of lead acid construction. They range from 0.7 Ah to 38 Ah and have sealing techniques which guarantee that no electrolyte leakage can occur. The batteries can be operated in any orientation and have an expected 5 years of service life. A stabilized constant charging voltage source is provided by the control panel with current limiting proportional to the batteries' Ah rating.

With the progression of microchip technology, which has led to more reliable and less expensive control panels, the electronic keypad is now widely replacing the mechanical keyswitch set up. In the case of the keypad the user enters a unique code to set the system rather than using a key to perform the 'on-off' function. Equally remote keypads (RKP) can be wired to the panel so that all operations can be performed from a remote point. Other keypad panels exist as stand-alone units and are wired directly to a power supply in a tamper protected case which has room to house the necessary standby rechargeable battery. Such power supplies can also be used to upgrade a battery panel installation or provide extra capacity when extending an existing installation.

Keypad panels have all the functions of keyswitch types but with a programmable code often of four digits. This can be changed by the operator or by the engineer. In some cases access to the inside of the panel must be made to do this. Switching off a zone is achieved by keying in an omit condition. A useful feature of some keypad panels is that only the first two digits of the code are used to set the system. This is an aid in cases where persons are required to set the system but it is not necessary that they know the full code for switch off purposes. Once set, all four digits must be entered to return to day or off. Entering a digit when the system is set may also start the entry time. This is valuable when an intruder could reach a panel and try entering digits in an attempt to randomly discover the code; only a short period should therefore exist for the correct code to be found. Anti-tamper circuits can be triggered if digits are entered in large numbers or for a greater time than is reasonable for legitimate operation. All control panels will use either individual light emitting diode (LED) displays or have a seven segment LED screen or a liquid crystal display (LCD) to show the system state and its zones. When remote signalling is used the system should latch and must be reset by the installer (engineer reset) or it may be reset by the user using an agreed procedure and after having contacted the central station. Otherwise the system is customer reset. If remote signalling is employed the audible devices may be delayed so that intruders are unaware that they have generated an alarm condition – this can provide an opportunity to apprehend offenders.

Advanced panels often incorporate internal speakers or sounders; others have a feature, during the entry cycle, that raises the entry tone if half the entry period has been exceeded. Let us assume that the total entry time was preset at 40 seconds. If the user on returning to the premises did not turn off the panel within 20 seconds then the internal speaker would raise its tone to give warning that a full alarm condition could soon occur. Twenty seconds after entering the building the external bell or siren would not sound, but after the full preset time it would. This is an excellent prewarning function and helps avoid unwanted external alarms occurring on the entry cycle, causing a nuisance to neighbours.

Still on the subject of advanced panels and special features there is a growing tendency for many keypad panels to have several user codes. These different codes, when keyed into the control panel, select different protection levels, in that they do not invoke every detector or zone that the full system incorporates. The term for the main code is 'master code' and for a lesser code the term 'cleaner code' is often used. In practice if one keyed the master code into the panel then the system would be either fully set or turned off in total. However, if the master code was used to set the system and the cleaner code is then entered only a portion of the system is switched off and access can only be gained to certain sections of a protected premises. This facility would allow a cleaner to perform work in specific areas, whilst denying access to other areas.

The employment of these different codes makes setting and unsetting easy and automatic without any need to omit or deselect zones on a day-to-day basis. These control panels, systems and detectors are wired in very much the same fashion as the simpler systems but will have been programmed in a particular manner to allow the use of the special purpose settings.

There are more special wiring systems used for the larger installation and the type which should be understood to some degree is the multiplex.

Multiplexed alarm systems allow the connection of many detection zones onto a single cable. These are individually identified on the control panel display face. Installation times are dramatically reduced using such a system, since all the data and power for every detector is carried on a single cable. For some systems different styles of transponder can be used to provide an interface between individual detection zones and the multiplexed cable run. Compact transponders can be located inside detectors. Larger cased transponders, which feature walk test LEDs, can link conventionally wired zones to the cable run. Alternatively they can operate with door contacts and detectors that do not have an inherent LED. The transponders are self-addressing and a specific detection zone is automatically determined by its position along the cable run, the transponder zone response being programmed at the control unit. Transponders may be scanned several times per second and only when a fault condition or alarm is sensed twice in succession will an alarm be made. Such scanning provides a reliable and secure communication function between the individual zones and the control panel. For the residential application, multiplexed wiring can combine intruder alarm, panic and fire detection in one system. Any combination of devices in different zones may be connected onto a single alarm loop as illustrated in Figure 1.8. Coding resistors are placed within the detection device and recognized by the control panel.

Although the multiplex method is intended for larger sized installations, there are developments in other intelligent systems to enable them to sense and then transmit the signals from the detectors to the control panel using fewer cables. These systems can even be used in domestic installations since the commissioning, servicing and future extensions can be quick and cost effective. Extensions are achieved by using plug-in addressable chips that are inserted in compatible detectors. These detectors extend across the sensor range in common use. Any normally open or closed detector can be used in conjunction with the discrete chip which is wired in parallel across the 4 core sensing line running back to the control panel. The cabling is 2 core for sensing only and 4 core for sensing and voltage supply to powered detectors.

It will be clear, therefore, that not every sensor needs to have its cable run back to the control panel but that it can pick up anywhere on the continually monitored sensing line. Different detectors can be mixed so, for instance, PIRs, PAs and contact switches may be on the same cable run.

There are other sophisticated system types and these can monitor the detection circuits by using end of line resistors (EOL), which prove if there are shorts or open circuits across the detection loops. The integrity of these resistor components determines the condition of the circuit.

Some insurance companies require event logs with printer facilities to give details of dates and events; data on system setting and unsetting may also be needed. Increasingly popular is the facility to upload and download which means that a panel's parameters can be passed from the service manager's office via a modem. Many of the latest advanced panels have a modem built in.

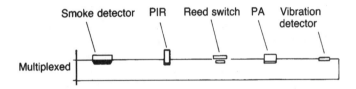

Figure 1.8 *Multiplex circuit*

Clearly the more technically advanced panels become, the more inherent the programming becomes but the cost of the components making up the system also increases. However, there will be a control panel and system to suit all sectors at an economic level and even with a budget panel many features can be offered. Options include 'silent PA' output, triggering remote signalling equipment only; 'double knock', preventing any detector causing an alarm unless activated twice within a short period of time (usually some 5 minutes or so); 'soak', enabling movement detectors or other powered devices to be monitored for stability when first installed. When a zone is selected for soak and the system is armed, any activation will be logged only and no alarm condition will be initiated. A 'terminator' option will enable the system to arm when an exit terminate button such as a bell push is pulsed closed, thus terminating any remaining exit time that the panel would have. Since the exit time can then be infinite it ensures that an exit fault cannot occur by exceeding the exit delay before pressing the button. Another feature often found is 'keyswitch set/unset'. This permits setting and switching off at a remote point by means of a key remote to the main control equipment.

We will now concentrate on the more standard control panel. A useful feature often incorporated is a fire zone which has a dedicated output to differentiate from the intruder signal. These panels have a clearly marked fire zone loop. The connection will be normally closed. They use 12 V powered smoke detectors with a relay output, hence they operate from the same supply as the intruder detectors. The control panel, however, will identify the fire detection separately.

At this point in time there is a huge choice of panels available. For the beginner wishing to install a small to medium domestic system, a suitable panel would be one with at least three alarm zones plus entry/exit and PA, with perhaps a fire zone as well. There are a good selection of units available to that end. The actual method of using the intruder alarm control panel and the means of setting and unsetting it depends very much on the panel being used and the way in which the detection system has been installed. If one or more remote keypads are employed then the system may be turned off from different points, but the setting operation will be found to be common.

Let us consider a particularly basic control panel and a system making use of a number of window/door contacts, pressure sensitive mats and motion detectors. The first step must be to interrogate the panel to ensure that the mains power is available – this will be indicated by an LED response. All protected windows or doors should be firmly closed and any objects cleared from above pressure mats. The panel can then be turned to its on or set condition using the key or appropriate code, or from a remote point in the protected area using a slave keyswitch or keypad. An audible tone will then be given at the panel by means of a buzzer or loudspeaker (LSO), or both as shown in Figure 1.7, and the correct setting of the system can be checked by reference to fault LEDs incorporated in the panel fascia. The user may then leave the premises by the designated exit/entry route, allowing the panel to become fully armed once the time delay has terminated. On re-entering the premises via the authorized entry point an entry tone will be given and at this stage the user must proceed directly to the control panel in order to return the system to the day or off condition. In certain cases different audible tones are given for entry and exit cycles and also for system faults. Some panels also allow the system to be set using an exit terminator rather than using a measured time delay. This is effected by a final setting switch employed on the final exit door or by a push button installed for this specific purpose.

The method used for dividing the premises and circuits into protected zones is given in the following section since, in the majority of systems, not every detection device is always required. The omit condition for unwanted zones may be achieved simply by selecting either a full or part guard position on the panel, or by keying in some form of omit procedure for keypad panels.

Developments for slimline microprocessor based control panels that are simple to install and easy to program are occurring rapidly and recent models feature a clear alpha-numeric display used for commissioning. Microprocessor based panels obviate the need for variant LED and audible tone responses by responding to simple questions and options to program, or the system may be set using an integral keypad. These control panels have a user friendly text display and record all events in a memory which may be called to the display at any time by either the installation engineer or user. These non-volatile memories (NVMs) store all programming data in case the control panel is ever totally de-powered of both mains and battery. When the user enters a preselected code a sequence of directions and questions used for testing or setting the system is initiated. Once the system has been completed its status will be displayed on the screen. One can certainly see a greater move towards quality in all aspects of intruder alarm components and in the advancement and diverse nature of the systems. Therefore it can never be possible to describe all of the different setting methods as they are very much governed by the precise system and the control panel in use. However, setting methods may become clearer once an understanding of how a protected premises is divided into zone circuits or areas has been gained.

1.3 Zones

The best way to visualize the simplest form of zoning is to consider a medium size house with a ground floor and first floor construction. If all the bedrooms were to be located on the first floor and the living, dining and kitchen areas were to be on the ground floor then the zones could be seen as Zone 1: ground floor and Zone 2: first floor. During sleeping hours the detection devices on Zone 1 could be made active, protecting the ground floor of the house throughout the night whilst the first floor would be switched off. This allows the upstairs of the house to be utilized at night but stops any intrusion downstairs while the occupants are asleep. If the premises were to be left totally unattended then both Zones 1 and 2 could be made active so full protection was afforded. In the case of a bungalow, the zones could be arranged so that the bedrooms and the toilet were on Zone 2 and hence left unarmed at night. In such contexts Zone 2 is known as a switchable zone i.e. it can be either on or off as it is independent of the rest of the system.

Let us look at the installation of a hardwired intruder detection system in an average size two floor home of perhaps semi-detached form. The house has three bedrooms which are all located upstairs. The occupant wishes for a cost effective system to be installed but want it to be simple to operate. Also the homeowner does not want overpowering security and feels the aesthetic value is quite important.

The system as planned on the ground floor uses one PIR detector to give a volumetric coverage, and four door contact switches. The PIR detector covers both the dining and lounge rooms (Figure 1.9). The first floor can be protected by arming certain doors that give access to rooms which contain valuables. Alternatively pressure mats could be used at appropriate positions on the upper landing.

A feature employed on many control panels is a 'walk through' on certain zones during the entry/exit period. This is often known as inhibit or as an intermediate entry route or circuit. It will be seen from Figure 1.10 how this is applied.

There are reasons why the control panel must be sited as shown. Providing that the user of the system enters the premises through an authorized route (i.e. the front door) then the PIR (which in this case protects Zone 1) is automatically inhibited. This enables the user to proceed to the control panel in order to return the system to the day

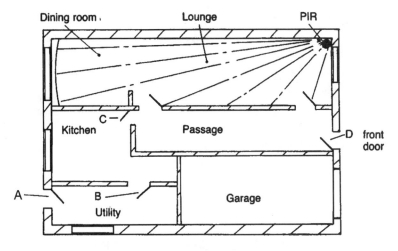

Doors A, B, C and D protected by reed contacts

Figure 1.9 *Zone planning*

condition. If an intruder were to gain access to the premises and trip the PIR without entering through the front door then the alarm would sound immediately without any entry delay. For protection of this nature a panel having either inhibit on Zone 1 during entry/exit periods or having an intermediate exit route (Ix) should be selected. In the latter case the PIR is simply wired into the Ix alarm loop. This is of course a far superior method of protecting the premises than

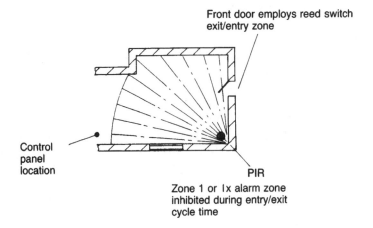

Front door employs reed switch exit/entry zone

Control panel location

PIR

Zone 1 or Ix alarm zone inhibited during entry/exit cycle time

Figure 1.10 *Intermediate entry on Zone 1*

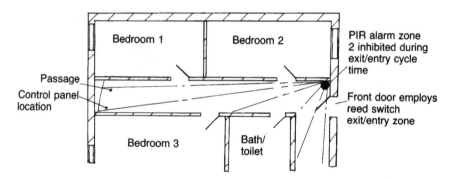

Figure 1.11   *Intermediate entry on Zone 2*

using a panel without such a function, since the PIR would then need to be wired in series with the front door alarm entry loop. If an intruder were to gain access to the house without coming through the front door then the PIR would only delay the system rather than signalling an immediate alarm condition.

Extending this practice, inhibit can also be used on switched zones with some panels. Let us look at a bungalow with a long corridor which adjoins a number of bedrooms.

In a bungalow of this form the corridor passageway is only to be protected when the house is not occupied since all the bedroom doors open on to it. Equally the bathroom/toilet doors must not be protected at night. Rather than use switches to protect all the doors the owner wants a PIR to be employed when the premises are unoccupied since he does not want to have to close all the doors when leaving the house. Therefore the PIR must be on a switchable zone, i.e. Zone 2. Since the movement detector could be triggered by the owner when entering the premises via the front door then the panel in this case must have a walk-through inhibit on Zone 2 during the entry period (Figure 1.11). In the case of a home that needs some bedroom doors protected at night because those rooms are not generally used, the installer may wish to put in a system that could allow those doors to be switched off at certain times.

Take for example a home with only one occupant who wishes the ground floor bedroom doors to be protected at night. However since visitors may sometimes wish to use those bedroom doors and the system is still required, a solution must be found for those periods.

One method is to use a panel with further zone capacity which can be omitted by the control panel even though the extra zones may be rarely used. An alternative practice is to put a switch in parallel with the detection devices that are to be switched off on certain occasions

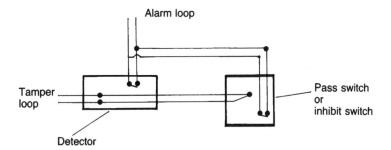

Figure 1.12   *Pass switch*

(Figure 1.12). This pass switch or inhibit switch must itself be anti-tamper, operable only by the keyholder of the premises.

Having now looked at how the essential zones can be either selected or omitted we should consider the result of a detection device being activated by an intruder when that device has been selected by the control panel and is in the 'on' condition, since an alarm condition will be generated.

1.4 Signalling equipment

Once the control panel goes into an alarm condition then it will provide power to the warning devices. These devices could be bells or electronic sirens, either of which can be mounted at various locations. In the main at least one internal sounder and one external sounder are used often supplemented by a visual warning device.

Xenon warning beacons or strobes with a high intensity light output are useful when a visual alarm indication is required over a wide distance, especially when a number of houses in close proximity possess intruder alarm systems. These units should be completely weather resistant (IP54, see Section 2.5) with the dome design allowing light to be transmitted across a wide angle. For security purposes beacons can be in many colours, although red is intended for fire alarm systems. They should also be capable of withstanding reasonable impact forces that could be applied from a distance by potential intruders or vandals.

Warning beacons are often mounted onto or inside bell boxes or outdoor siren boxes so that they are used in conjunction with the audible output and can be wired in parallel with the internal sounder terminals. Many control panels have special strobe outputs which can remain active with the internal sounder when the external audible system has cut off at the prescribed bell duration time. In such cases the strobe will continue to give its visual signal even though the external signalling device has shut down, and it will continue to do so until the control panel has been switched off by an authorized party.

There is a growing tendency to connect communicators or speech diallers to alarm panels to relay messages to a number of personally programmed telephone numbers when the alarm has been triggered. These are reliable as they do not use a tape mechanism but employ a voice recording stored in a battery backed RAM, the personalized message having been recorded by the user. These communicators are mains powered, battery supported and easily wired to any intruder alarm control panel. They accept either a 12 V trigger or sometimes an opening or closing circuit and continue to dial the pre-programmed numbers until they receive a response, i.e. until the call is accepted on a dialled number. The diallers are plugged into a standard telephone socket by a double adaptor. Other dialling practices are discussed later in Section 5.5.

Modern security installations often use multi-sirens which replace the internal siren and entry/exit warning sounder since both functions are accommodated in one compact unit. The high noise output frequency tone and a reduced noise output monotone are produced by independent solid state drive units.

It is always important that the entry/exit warning buzzer tone can be heard at all points that can be used for entering and leaving the premises. The multi-siren can therefore be situated away from the control panel. Many panels have outputs for extension speakers which serve the same purpose and can be placed at a number of locations, also warning an intruder that an alarm condition has occurred.

One will appreciate that the control panel, having generated an alarm, will have achieved its task by causing the intruder to leave the premises since the alarm has become a centre of attraction. However, much distress can be caused to neighbours by an alarm sounder after a burglary if houses are in close proximity to one another. In the event of a keyholder not being readily available such an annoyance can continue for many hours, hence the need for alarms to shut down after a prescribed period.

Many intruder alarm control panels do automatically reset the control panel after an alarm, but in order to do so must isolate a part of the system, i.e. the offending detector if still in an alarm condition. An example of this is a door left in an open position.

Naturally the level of security becomes impaired and in theory an intruder could elect to enter the premises once again. In order to overcome this problem 'auto-reset/bell shut-off modules' can be adopted. These modules are designed to interrogate the control panel at the reset time. If an alarm is generated the module permits the panel to operate the alarm sounders. However after the preset alarm period the control panel is interrogated by the module and if still in a fault condition it will inhibit the exterior alarm but retain the internal sounder.

At intervals the module will continue to interrogate the system and when the alarm circuit has become clear it will silence the interior sounder and reset the alarm system. All the keyholder need then do is close the door which caused the original activation. It should be noted that space detectors would tend to reset after an alarm unless they had become defective in the process.

It will be seen that with intruder detection devices that possess normally closed alarm contacts when set, that an alarm condition would exist if an open circuit were to occur, for example if wires were to be cut. The external sounder would then derive its power supply from the control panel to generate the alarm condition. However if an intruder were to cut all the wires in the vicinity of the control panel and the wires which form the alarm loop, the wires supplying power to the external sounder could also be destroyed. No alarm could then be given. For this reason external warning devices should be provided with their own standby power in the event of this happening. In practice this is done by enclosing within the bell box a self-activating bell mechanism often referred to as a self-activating bell (SAB). Figure 1.7 shows the typical wiring for an SAB. The unit contains the necessary electronics and nickel–cadmium battery to form an independent power source in the event of wires to the external warning device being cut. This is effected by running four wires from the control panel up to the external sounder rather than two. The electronics package providing this function should also be equipped with an anti-tamper switch to prevent any attempt to remove the cover surrounding the sounder. If such a switch is operated whilst the panel is set it should generate a full alarm. If the system is in the off or day position the panel should preferably only give an internal sound to indicate a tamper condition.

SABs should always provide a four wire control connection. Universal types can satisfy a positive or negative switched bell input with positive or negative tamper return. A 20-minute bell ring should be inherent. The nickel–cadmium battery is trickle charged from the hold off supply from the control panel, hence the bell will only ring from the ni–cad if this supply were to fail. The recharge time is typically 24 hours and the temperature range from –20°C to 50°C. After manufacture the unit is generally protected with a conformal coating to give a high degree of resistance to moisture but further protection should be added following installation. Attention should be given to the terminal block and surrounding area. A flexible acrylic coating of silicone resin or a protective lacquer can be used for this purpose.

There are many modern slimline bell box sounders on the market, in various colours, complete with strobes, sounders and SAB devices. Often these are louvreless, of double skin design and with standard drill protection which all prevent the injection of sound deadening foam into the unit. Charge LEDs can act as visual deterrents and adjustable cut off timers ensure that the unit can only stay in alarm for a prescribed time.

Any alarm system needs visible bell boxes for their deterrent value, but in many cases if they are very accessible it can pay to hide additional sounders behind air bricks, under eaves or in signs if one suspects that the master unit is particularly vulnerable. These slave units can of course also have SAB support. For high risk applications high security sounders may be used. These do not derive their power from the main control panel, but have an integral battery. A relay in the bell box is triggered by the control panel when in alarm and this actuates the battery which itself has a charging circuit. Additionally, for either high or low risk installations dummy deterrent boxes can be mounted on the protected premises to promote security. Dummy boxes can also have LED indicators working from a simple self-contained module powered by a 1.5 V battery and these are becoming increasingly popular.

Having gained an insight into the array of available bell boxes it is important to look once again at the distress that can be caused to neighbours if an alarm fails to shut down after a reasonable period of time. All new panels must have a maximum shut down of 20 minutes, whilst for older panels a module can be easily interfaced between the panel sounder terminals and the SAB to close down the output. However if an SAB was to lose its supply, i.e. the 12 V across points A and B in Figure 1.13, then the charged ni–cad standby battery could supply power to the sounder for well over the 20-minute period. This supply could have been lost due to mains failure and then flattening of the main control panel battery. Equally a break in wires A or B would create the same effect. The solution is to install a special purpose SAB which incorporates a timer that disconnects the standby battery at a selected period of up to 20 minutes. The wiring is identical to that of its non-timed counterpart.

The 20-minute shut-off value is quoted by the Department of the Environment in Section 4.1 of the 'Code of Practice on Noise from Audible Intruder Alarms' and in the 'Control of Pollution Act Part III

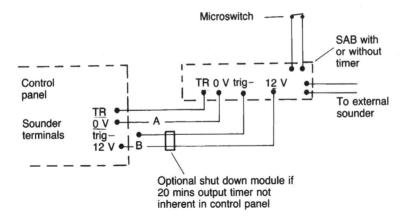

Figure 1.13 *SAB wiring guide*

– Noise'. This is invoked in the Environmental Protection Act Section 80 which classes noise from an alarm outside of the detailed time limit as a statutory offence and subject on conviction to a heavy fine.

We mentioned at an earlier stage the use of high security sounders and the fact that they do not derive their power from the main control panel when in alarm but use the energy of the integral battery. This is to allow the use of higher output devices and greater levels of sound. They are referred to as self-contained boards (SCB) and are used in high risk situations. In reality, to complement these devices we shall see a greater use of remote signalling which, when connected to central stations, will be interfaced by either a digital communicator or, for RedCARE networks, by a subscriber's terminal unit (STU).

The only other consideration is that of appreciating the different types of line that are in common use to transmit the alarm signal. These are simple telephone company voice-grade lines not monitored for disconnection.

- The direct line (individual). A single line running direct to the police or the central station. It is direct in the sense that it is not linked with an exchange. It is designed to be used with a communication device that itself can be monitored for line failure.
- The shared direct line. As for the individual direct line but carries signals from a number of other protected premises in the same general area. These signals are multiplexed on to the line being carried simultaneously.
- The indirect non-dedicated exchange line. This is indirect since it is directed via a telephone exchange so that a switching process exists. It is the common method adopted for mid-level security risks.

● The dedicated exchange line. Similar to the previous line but is dedicated to the alarm and not used for other outgoing calls. It is not as cost-effective because of the restriction of the line to the alarm output signal.

The alternative practice to using these lines and a communication device is to use the RedCARE technique or Paknet which are discussed in Section 5.5 which deals with 24-hour central stations.

Before we leave the subject of signalling equipment we should say that the control panel itself should always be placed in as secure a position as possible if it could be subject to forced attack with resultant damage to its signalling potential. It is always advisable to use an RKP (remote key pad) if the system must be operated from an exposed point, with the power supply installed in a hidden location. This is far better for security because it ensures that the heart of the signalling system is not accessible and if the RKP were to be attacked then the signalling would not be affected. RKPs, being so much smaller than a power supply or regular size control panel, are more aesthetically pleasing, and hence are not out of place when sited in an area that can be viewed by the public. Equally they can be on display at the main entrance in a domestic application. There are also many RKPs and control panels that allow certain detection zones to be programmed for 'chime'. This function can be used, for instance, on a front door which when opened will cause the panel or RKP to emit a tone somewhat similar to a doorbell. Such a feature can announce the presence of a person entering either a shop or a house and can also be adopted as a means of giving additional security to the control panel.

Uploading. Downloading

While dealing with the subject of signalling equipment we should overview the techniques of uploading and downloading since they are related to remote signalling methods.

Systems that send an alarm signal to private security central stations are well recognized within the security industry for most types of risk. The growth of their facilities is a measure of the advantages that remote signalling can achieve. It optimizes police time and resources and allows the receiving centre operators to make value judgements and to interrogate systems.

It is now the case that in the central station the software is seen as an expert system for handling data in order to determine the responses that are then to be made. Instructions to humans and response personnel can be made available at an early stage. Software programs must, of course, be both fast and simple to operate yet have inherent flexibility. One of the new features that the software must now possess is uploading and downloading protocols. This is effectively the process of allowing a security system to be remotely programmed or interrogated via the telephone line and a computer. The ability to programme security systems from a remote location rather than from the installation itself offers tremendous advantages to the installer and end user. Although the central 24-hour station will have the required programme capacity to perform these functions, even the more basic PC can carry out the task once connected to the telephone line via a modem.

The knowledge that assistance is only a telephone call away is very reassuring for all end users. If requirements change, the system can be easily reprogrammed remotely to allow for such things as addition of new keyholders, change of use of an area or extension of setting times and methods.

It is currently the case that uploading and downloading are enhancing communication. A plug-in or boxed digi modem is connected to the control panel and allows remote interrogation to be carried out via the computer network. This digi modem can also act as a signalling device to generate a standard alarm activation signal to the monitoring station.

In practice the engineer must first obtain confirmation of the compatibility of the software packages and the control equipment as, in the main, digi modems are used with control equipment of the same manufacture.

As an extension of this concept we can see how a number of independent software modules can be adopted and monitored at a PC-based control centre to cover a number of systems in different sites. For instance, the head office of a bank can monitor all the systems of its branches, the operation being controlled by detailed graphical representations based on Windows technology with Help facilities. The programming can also include methods of alarm confirmation in accordance with the ACPO Policy such as Point ID and alarm verification. It must also include zones or wards which are, in effect, detector groups.

One will recall from Section 1.3 how zones are established and, at that stage, we determined a simple zone strategy. If we now take this a stage further it is possible to visualize an application in which it is requested that a commercial premises is ascribed zones made up of three essential areas which are themselves made up of a number of circuits. In these premises the three zones are:

● Zone A – sales department and canteen.

- Zone B – workshop and canteen.
- Zone C – stores and canteen.

Table 1.1   *Zones/wards ascribed to a number of circuits*

Circuit No.	Location	Zone A	Zone B	Zone C
01	Sales entrance door	✓		
02	Sales PIR	✓		
03	Workshop PIR (1)		✓	
04	Workshop entrance door		✓	
05	Workshop PIR (2)		✓	
06	Workshop internal door		✓	
07	Workshop PIR (3)		✓	
08	Sales internal door	✓		
09	Canteen door	✓	✓	✓
10	Canteen PIR	✓	✓	✓
11	Sales entrance door			✓
12	Stores PIR			✓
13	Stores internal door			✓
14	Sales (office PIR)	✓		
15	Service entrance door			
16	Service PIR			

Table 1.1 shows a zone structure in which groups of detection circuits may be set or unset independently of each other. From this the detector circuit numbers are programmed into the part-set functions so that during the setting procedure the user at the control panel need only further select an A, B or C button to arm the system and determine the extent to which it must be active. As an example, if the system is turned on and then button B is selected, the system that becomes active will be the workshop and canteen, i.e. detectors that are wired on circuits 03–10 with the exception of the sales internal door. Circuits 15 and 16 are separately selected as need be at the control panel. If no part-set buttons A, B or C are pressed then the full system is selected, i.e. circuits 01–16.

From the example we can appreciate how these part-set conditions can be used in any system from a simple domestic installation to a complex commercial or industrial application. If the system is also equipped with upload and download facilities then, from a remote point, all system attributes can be managed, and sites with individual buildings or branches can be set and interrogated from one control house. Customer histories can also be maintained on a system hard disk with encryption communications supported over the telephone network.

We noted earlier that programming can also include alarm verification and Point ID reporting in accordance with the ACPO Policy. Such systems are to confirm, to as high a degree as possible, to the central station that an alarm activity has been caused by a genuine intrusion. They may be verified in a visual, audible or sequential form. The latter is the least expensive and easiest to install. It uses an increased number of existing standard detectors employed within each zone of the protected premises. An alarm is only confirmed after two or more sensors have been activated in succession within the confines of one protected zone. Point ID is a method widely used in North America and of most benefit in larger premises. It is an integral part of sequential confirmation and the transfer of all information from the security system to the receiving station. Using high speed communication every alarm that has been generated can be uniquely identified by its type and zone among other reporting data.

When integrated with any remote signalling medium the security system will also be able to differentiate between signals given from automatic detectors and those generated by a deliberately operated device. It is the latter we look at next.

1.5 Personal security

Panic buttons or Personal Attack (PA) deliberately operated devices can be used to trigger an alarm by pressing a button on the unit if an occupier is threatened. Equally they can be used to sound the exterior sounder if a person becomes ill or needs attention.

PA devices are active for 24 hours. Most modern control panels have separate PA outputs and indication. If no independent output is available then they can be wired into a 24-hour tamper loop which provides full alarm output at all material times. PA devices are generally sited near the bed head in the main bedroom and at the front entrance to the premises. However, other locations may be used to suit one's specific needs.

The alarm contacts must be wired in series when there is more than one device used on a given circuit. They should also mechanically latch and require a positive setting method; often this is achieved by using a key provided with the units. Momentary or self-resetting units can be used providing the control panel has an LED indicator for the PA function.

Once activated the alarm should only be capable of being cancelled by performing a resetting operation at the control panel, that is turning the panel 'on' and then 'off' for keyswitch panels or entering the user code into keypad operated panels.

More sophisticated panic button modules feature two buttons mounted behind a tactile keypad and have a preset, pre-timed action capable of operating in a number of phases. Pressing one of the buttons will set the device into an automatic time delay mode. Failure to press the other button within a preset time will cause the device to activate therefore triggering whatever apparatus the device is connected to. Pressing this button within the time limit cancels the timer and restores the unit to its standby mode. The obvious value of the automatic timed mode is, for instance, if an occupant is attacked before being able to press the attack button on a conventional PA. With the automatic timed mode the alarm would sound if the cancellation were not received. An LED is provided to give the status of the PA, hence the occupant should not be able to forget to cancel the device from its automatic setting. This indication could be:

- Standby   LED off
- Timing   LED flashing
- Activated   LED on standby

If one wished to cause an instant alarm then pushing both buttons simultaneously would initiate immediate operation irrespective of the mode of the PA.

Advantages of this form of PA, which is powered from the control panel, is that there are no resetting keys to lose as the device features a touch button reset. Also the operation of both buttons for immediate use greatly reduces the risk of false alarms.

A homeowner, on going to answer a call at the door, would press one of the buttons to set the device into automatic time delay mode, but if he or she were attacked and unable to return to the unit within the prescribed period then the alarm activation would occur.

Clearly the PA unit forms an essential part of any intruder alarm or security system since crime statistics indicate that greater violence is being used against persons attempting to protect their property. Although the device itself usually has a reed switch some versions do feature precision snap action microswitches to perform the internal operating function. Although most attack devices are intended to be hand operated more specialized versions are available for activation by foot, being mounted slightly higher than floor level in a hidden location.

The standard hand operated button type is time honoured and used to great effect close to counters and tills, not only in banks but also in the vast majority of shops and retail outlets. They are often found in a horizontally mounted orientation under the counter where they can be reached from the position one would stand at to operate the till. When used in a system that has automatic dialling equipment installed, the communicator will have a dedicated channel for PA so that the recipient of a call will be able to act in the most appropriate manner.

At a later stage in the book Figure 2.13 shows how a PA push button can be wired to produce an audible and visual output by means of a 12 V power supply and relay; the lighting is mains power derived. Equally in Section 5.3 we learn that portable pocket sized PAs can be carried for use with wireless systems and also how they may be integrated into existing hardwired circuitry.

It can be seen that it is vital that attack devices do not false alarm and fortunately in practice the incidence of false alarms on PA loops is extremely low.

An option on some control panels is a PA output which is activated if two designated keys are pressed simultaneously. This is a welcome feature if the panel is close to the front door and can be reached easily in the event of a forced attack being made.

A further option on certain control equipment is a duress code. This provides a silent signal over the remote signalling network if a certain code is entered rather than the correct operator code. It can be used to turn off a system but, unknown to an intruder, it also alerts the central station that a duress condition exists.

1.6 False alarms

In real terms false alarms form a major subject but in practice the responsibility for limiting and ultimately removing the incidence of false alarms rest with the system operator, the installer and the manufacturer of the goods.

It is clear that if an intruder alarm system is to retain its credibility then it must not falsely activate. Well-designed and correctly sited components should alleviate the problem but certain other factors must also be realized.

Alarm circuits are affected by dry joints that can occur, by terminals becoming loose or by corrosion and high resistance over the contacts. These will all cause false alarms. Voltage drops caused by long cable runs also causes problems. More power supplies or larger power supplies can help in this case even though with the present generation of electronic components power consumption is considerably less than it was some ten years or so ago. In general once a system is installed the resistance of all circuits should be recorded. This should then be verified periodically. If a reading alters it should be immediately investigated.

Radio frequency interference (RFI) from radio transmitters or from electrical devices and power lines can also cause false alarms by activating a detection device. The possibility of activating alarm equipment by RFI is well known to alarm manufacturers who are responsible for the careful design and utilization of components to suppress unwanted signals.

Unfortunately detection devices are prone to RFI because the connecting cables serve as antennae for the energy. During installation therefore it is important that all cables are firmly terminated and loops not formed within detector housings. In other cases appliances causing electrical arc interference must be isolated. Purges purpose built into spurs can be used to this end and simply replace the control panel fused spur.

With regard to detection devices magnetic reed switches are rarely troublesome and, provided that they are terminated with care, their performance should be consistently good. However for maximum reliability and improved performance rhodium plated switch contacts can be used. These may be selected for ease of installation with maximum gap tolerance to reduce false activations. One hundred per cent product inspection is available on these products to full specifications. To satisfy their full market potential both flush and surface versions are available with different gap settings. Adjustable plunger switches and overhead door contacts in metallic enclosures can also be obtained. Roller shutter contacts in either aluminium or plastic housings can be employed on large or badly fitting shutters or doors because they feature a more powerful magnet and can cater for movement greater than normal or for misalignment.

Having installed reed switches it only remains to ensure that the gap between the switch and magnet is within tolerance and the door or window/shutter should be opened several times after commissioning to prove that it cannot go open circuit.

Alongside the reed switch PIRs are used in great volume and their limitations should be understood. The main false alarm sources of PIRs are

● Sunlight shining directly onto the sensor.
● Strong air draughts onto the sensor.
● Animals and pets entering the zone.
● Heat sources in a zone.

Sunlight directed onto the sensor is always a potential hazard. However strong sunlight entering a room and rapidly heating up an object in the field of view of the PIR can of itself present problems especially if sited close to the sensor. Strong air draughts, especially if associated with a heating effect, are also extremely troublesome.

Animals and pets are always liable to cause problems where PIRs are used but can often be catered for by using purpose designed detectors with flat beams or flat fan coverage. This version tends to be mounted at a height of roughly 1.5 m and to give 0° declination. In theory pets and animals tend to move under the zone coverage but one should be aware of objects such as chairs that animals can jump onto with obvious consequences. Certain PIRs, rather than being purpose built to avoid pets, are supplied with blanking pieces which are fitted to the inside bottom section of the standard volumetric lens.

Open fires and boilers are extremely troublesome and care should be taken to avoid PIRs monitoring the heat produced by these objects. The blanking pieces previously mentioned can also be placed over any other part of the lens to stop these particular areas being viewed. The area still in view can be easily checked by watching the operation of the walk test LED which is incorporated in every PIR. It must be said that PIRs are capable of exceptional detection with high immunity to false alarms provided that they are sited correctly. Also after siting and wiring all knockouts must be sealed to ensure that insects cannot enter the housing.

Having accepted the limitations on the siting of PIRs the installer must then select the appropriate device to suit the application. Mention has already been made of separate dual elements being incorporated within the sensor to give balanced detection and excellent immunity to false triggering. In these one element is designed to produce a positive voltage on receiving heat and the other a negative voltage, therefore if both elements receive heat simultaneously the voltages cancel out and no alarm is given. However the design is such that the energy from an intruder must affect only one sensor element at a time to generate the alarm voltage, hence movement must occur across the detector. Therefore changes in ambient temperature, acoustic noise or sunlight should affect both elements at the same time and hence be cancelled out.

As an extension of this technology quad element detectors became available. They are known as PIR Quads and have two dual elements. The two outputs are fed to a signal processing unit which will go into alarm only when two signals from the quad system exceed a predetermined threshold level. It becomes apparent that in the first instance quality detection and control equipment must be specified in order to avoid unwanted activations. Also, detectors such as those that recognize vibration or impact must have their sensitivity adjusted to a sensible level. If necessary devices that employ different detection techniques within a common housing, such as combined or dual technology, must be installed in any environment where a false alarm hazard exists.

Mention has already been made of certain detectors that have a first to latch facility because they only give a momentary pulse when in alarm, and it is important that their identification LED latches so it is known which detector caused the alarm signal. If this method is not employed it is vitally important that the control panel does have an LED that will itself latch to identify the zone on which the alarm originated and hence the detector. It is bad policy to have multiple detectors on one zone with no method of establishing the alarm source or using a control panel without a latching LED to achieve the same end. In such cases it can be extremely difficult to trace and solve false alarms.

As a further step to ensuring quality, many organizations having had a security system installed want a maintenance contract in order to have an ongoing service and a high level of system reliability. This is a requirement of some insurance policies. Maintenance contracts are vital for alarms which go beyond audible outputs to remote signalling. Originally many contracts were the equivalent of full rental agreements, in which the customer never actually purchased the equipment itself but paid only for the use of it within the rental period. The rental charge was at a fixed rate so it was always in the interests of the supplier to maintain the equipment well in order to keep down any costs arising from a call-out service. More recent arrangements see equipment connected to telephone lines only being rented, whilst the remainder of the equipment is purchased outright by the customer. Within the marketplace there is always some agreement that can find favour with any customer and installer to ensure that systems are essentially reliable. Small shop-type installations with only a local alarm sounder may well want some form of service contract with perhaps a separate arrangement if any call out is needed.

If all of the practices we have now looked at were followed and essential maintenance was provided then the incidence of false alarms would surely be at a tolerable level. Equally false activations could be more readily traced to reduce this incidence. Also if the installation was in accordance with recognized codes of practice and standards then unwanted alarms would be virtually non-existent.

This was addressed by the Association of Chief Police Officers (ACPO) Policy on police response to alarms. It states: 'The aim of this policy is to enable the police to provide an effective response to genuine intruder alarm activations, thereby leading to the arrest of offenders and a reduction of losses by improving the effectiveness of alarm systems and reducing the number of false calls to the police'.

The Policy is quite rigid in most of its requirements but it only applies to those systems with remote signalling where there is a police response. Under the Policy the police will only respond to alarm systems which have a unique reference number (URN) and for a company to be able to issue URNs they must be recognized by an Inspectorate Body.

There are two types of signalling – Type A and Type B:

- Type A. This is remote signalling to a 24-hour central station or the police direct and has a URN.
- Type B. This includes audible-only alarms, automatic dialling alarms, alarms installed by non-compliant companies or monitored by non-compliant central stations or alarm receiving centres, and does not have a URN. The police will only respond to this type if a third party can confirm an intrusion is taking place.

There are three levels of police response under the Policy:

- Level 1. Immediate.
- Level 2. Although immediate is desirable, other situations may take priority.
- Level 3. No guarantee of a response. Keyholder to investigate.

It can be noted that any system with a URN is level 1.

If false alarms occur four times in a 12-month period then the police response will be lowered as follows:

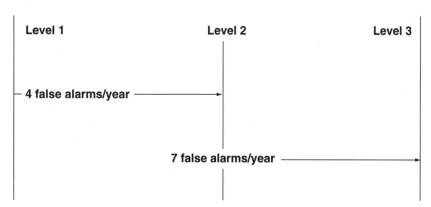

Under certain circumstances URNs can be withdrawn. Under the Policy there are particular needs for PAs which must be separately identified. With regard to false alarm prevention, emphasis is placed on users, installers and central stations to filter out false activations. It is said that the alarm receiving centres, operators at these centres and the installers are independently inspected to comply with Industry Standard Codes of Practice.

There is also a need to have remote signalling featuring either an open/close or abort signal function to be able to indicate to the central station that a system is set or unset or has been misoperated. This is to remove false alarms caused by operators setting and unsetting a system wrongly. With this in mind, police forces will set a threshold level for false alarms per system referred to as FASA.

This leads us to the recognized codes of practice and standards that govern intruder alarms.

1.7 Standards. Codes of Practice

Due to the growth of the security industry over the years it was always felt that a self-regulatory body was required which would set standards and act as a responsible representative for this sector of industry as a whole.

It was for this reason that the British Security Industry Association (BSIA) was instrumental, in conjunction with the insurance industry, in setting up the National Supervisory Council for Intruder Alarms (NSCIA).

The principal task of the NSCIA was to carry out inspections of all equipment installed by member companies to ensure that they met with British Standard 4737 'Intruder Alarm Systems in Buildings'. The original NSCIA has now been superseded by the National Approval Council for Security Systems (NACOSS) who hold a list of Recognized Firms in order to give reassurance as to the quality of alarm installation and maintenance. NACOSS is now the industry governing body and the most influential Inspectorate Body. There are, however, other Inspectorates that fall within the ACPO Policy to include the Security Systems and Alarms Inspection Board (SSAIB), the Alarms Inspectorate and Security Council (AISC), Integrity 2000 and the Independent Alarm Inspectorate (IAI). The adoption of these Inspectorate Bodies within the ACPO Policy widens the choice of organizations that an installer can join in the event of him needing to be able to obtain URNs.

In the first instance manufacturers of intruder alarm systems and components need guidelines to which they must work to ensure that the products they make satisfy a fitness for purpose. To this end British Standards are written in broad terms to produce a document that all competing manufacturers can use as a base for their goods.

BS 4737, which exists in many parts, can be used to establish parameters for products depending upon that product's precise nature. However since this standard extends beyond product manufacture guidelines to embrace also installation practices, it has come to form the base document used in the intruder alarm industry. Since the reader may also come upon the term BS 5750 (also known as BS EN ISO 9000/1/2), relating in one form or another to manufacture, it is correct to explain why this British Standard may also at times be invoked.

BS 5750 'Quality Systems' is the UK national standard for quality. This makes no reference whatsoever to products of any kind but it specifies the basic requirements for an adequate quality system and the way a company's business activities must be organized. In effect it has been written so that it may apply to any form of industrial or commercial operation. In all cases BS 5750 gives the requirements to be met during the operations which lead to a complete product. It exists in numerous parts and a manufacturer must first establish the part relevant to his/her area. If a manufacturer is assessed and found to have practices in compliance with the relevant requirements then that manufacturer can confidently promote the business and quality levels of his/her products. If a manufacturer's claims to BS 5750 can be complemented by his/her product complying with the necessary part or parts of BS 4737 then this provides a high level of confidence in the product to the purchaser. The manufacturer will be seen to have had his/her production methods and test and quality assurance facilities judged for capable and constant reproduction of the component or systems.

In the British domestic appliance sector products for use in the home are often subject to approval by an independent party. That is they are independently tested and approved for safety and durability of safety in accordance with the relevant British Standard. This proves their level of manufacturing quality and ensures that the rated parameters are still being met after a prescribed number of operations. Such schemes work well because it is clear that these electrical appliances are often operated by persons unclear of electrical technology and safety must be inherent.

In the USA Underwriters Laboratories (UL) operates similar schemes that also extend into the security manufacturing industry. However UL has strong links with insurance companies. It was founded in 1894 as an independent non-profit-making corporation to test products for public safety, and their service is used by insurance companies as a means of assessing the risks to be covered when providing policies. The majority of manufacturers of electrical products who wish to market their goods in the USA seek approval as a safeguard against legal claims in the event of any accident arising from the use of their products. Indeed, approval in many parts of the USA is now mandatory for all classes of non-military electrical goods. For this reason one may see security industry components claiming to have been tested by UL, or by VDS in Germany for the same reason.

British Standards however have not been endorsed by Parliament and approval by an independent party is never compulsory. Nevertheless, manufacturers could always submit their products for type testing with a view to gaining an appropriate certification report to obtain independent evidence of compliance. It is essential in all ways to recognize British Standards as valuable documents and BS 4737 in particular when looking at any aspect of intruder alarms. Part 1 of BS 4737 gives assistance to subscribers, alarm installation companies, insurers and the police in order to achieve a completely accurate specification of the required protection for audible or remote signalling systems. Part 2 refers to deliberate operation systems including hold-up and bandit alarms whilst Part 3, which is in various sections, covers the diverse component specifications and continuous wiring used as detection devices. The code of practice for the planning installation of intruder and deliberately operated systems is given in one section of Part 4 to cover those systems constituting Parts 1 and 2. Maintenance and records, and action to be taken in the event of false alarms, are given in a further section of Part 4.

In essence it would be incorrect for this book to attempt to detail specific requirements for each application, as once the system has been determined for particular premises various clauses from different Parts and Sections of BS 4737 must be invoked to ensure that all necessary details are comprehensively met.

BS 4737 is well recognized within the security industry and it has some alignment with International Publication IEC 328 produced by the International Electrotechnical Commission (IEC). Certain other relevant British Standards exist and the reader should be aware of them. BS 6800 'Specification for Home and Personal Security Devices' details performance criteria for simple security devices that give an indication of warning when activated. Typical types are video recorder protection devices, door chain alarms, access control alarms, movement detectors, door glass-break sensors, pressure mats, random switching devices, light sensitive devices and programmable switching devices. Many of these products do gain a mention in other chapters of this book. All of these devices are intended for use separately as individual items and not as part of interconnected systems. This British Standard appears with the object of providing users with a product that has an acceptable level of security consistent with economically priced products and to guide manufacturers towards the consumers, expectations of such products. BS 6707 'Specification for Intruder Alarm Systems for Consumer Installation' caters for self-contained units, kits and PA units intended for the 'do-it-yourself' customer. It gives advice for the planning, installation and use of a system purchased in kit form. This kit must contain sensors and detectors, warning devices, control equipment with tamper detection and circuits or circuitry. Battery only systems must incorporate a voltage test facility or battery state indicator. Self-contained units may have all of these in one enclosure.

It follows that a British Standard does exist for the differing levels of installation and product types.

Before concluding the subject of intruder alarms we should look back to the professional level BS 4737 and summarize Section 4.2 which is that part of the standard giving recommendations for the preventative and corrective maintenance and the keeping of records. If the installer follows this in total then the level of quality should remain high.

'Preventative' is defined as the routine servicing of a system carried out on a scheduled basis, and 'corrective' as emergency servicing in response to the development of a fault. For a system with local audible signalling preventative maintenance should be made annually; for systems with remote signalling twice yearly will be necessary. Checks should essentially cover the installation, location and siting of all equipment against the system record and the satisfactory operation of each sensor, detector and processor. All flexible connections should be tested together with power supplies and control equipment. In addition a full system operational test should be performed with attention being paid to the operation of all audible and signalling devices. A log should be made of the system of records and this should be protected from unauthorized access. Historical records must include dates of any visit to an installation with faults found and remedial action. Any alarm call should be recorded with details of action taken and, if known, the cause.

In the event of a false alarm the following must be checked.

● The subscriber's operational procedure to ensure that all users are conversant with the system.
● The premises for a change of use or structural alterations or a change in electrical supplies that could affect the alarm.
● Any possible source of environmental interference, e.g. heating systems, automatic lighting control and radio frequency interference.
● The correct operation of the system to include all interconnecting wiring and tamper circuits.

In conclusion we can say that installers and manufacturers alike must show a responsible attitude to the particular British or International Standard being invoked whether it be for the professional or do-it-yourself sector in order to achieve a high level of credibility in the industry. To this end British Standards are a consensus of opinion formed from the opinions of many authorities, including manufacturers and consumer bodies. Individuals thinking it is safe to deviate from the authorities' route will inevitably bring about a move towards shoddy goods and installation practices and a decline in the industry as a whole.

To complete this chapter we can refer to the BS handover checklist, other essential Standards, Codes of Practice and further sources of information.

Handover checklist for an intruder alarm: BS 4737

- The installation is to be in accordance with the Specification, to comply with BS 4737 and to be a high standard.
- The subscriber premises are to be left tidy.
- Detection circuits are clearly indicated.
- Detection circuit insulation and continuity/resistance measurements are logged.
- The mains connection is permanent (not plug and socket).
- Supply fuses are correctly rated.
- Batteries are marked with the date of installation.
- Dc current loading of power suppliers are recorded.
- On removal of the mains supply the battery voltage of all equipment is within specified limits and the system operates normally.
- There is adequate standby capacity to meet BS 4737.
- Check the audible alarm on activation and that SAB or SCB devices work correctly.
- Remote signalling apparatus gives correct transmission.
- Engineer-only reset facilities are available as applicable.
- All tamper detection circuits operate correctly.
- All detectors give correct operation at the control units.
- Check the area or volume of movement/vibration detectors.
- Check beam interruption detectors for correct alignment.
- Ensure that there is a system operating procedure displayed near to the control unit.
- Confirm the exit/entry route time delay if used and record the periods.
- Set the system. Operate a detector and check the signalling.
- Show the subscriber the extent of the protection and the correct system operation.

● Check the documentation against BS 4737.
● Remove all surplus and waste from the site.

Governing British Standards

BS 4737: Part 1: 1986
Intruder alarm systems in buildings. Specification for installed systems with local audible and/or remote signalling.

BS 4737: Part 2: 1986
Specification for installed systems for deliberate operation.

BS 4737: Part 3: Section 3.0: 1988
Specifications for components. General requirements.

BS 4737: Part 3: Section 3.1: 1977
Requirements for detection devices. Continuous wiring.

BS 4737: Part 3: Section 3.2: 1977
Requirements for detection devices. Foil on glass.

BS 4737: Part 3: Section 3.3: 1977
Requirements for detection devices. Protective switches.

BS 4737: Part 3: Section 3.4: 1978
Specifications for components. Radio wave Doppler detectors.

BS 4737: Part 3: Section 3.5: 1978
Specifications for components. Ultrasonic movement detectors.

BS: 4737: Part 3: Section 3.6: 1978
Requirements for detection devices. Acoustic detectors.

BS 4737: Part 3: Section 3.7: 1978
Requirements for detection devices. Passive infra-red detectors.

BS 4737: Part 3: Section 3.8: 1978
Requirements for detection devices. Volumetric capacitive detectors.

BS 4737: Part 3: Section 3.9: 1978
Requirements for detection devices. Pressure mats.

BS 4737: Part 3: Section 3.10: 1978
Requirements for detection devices. Vibration detectors.

BS 4737: Part 3: Section 3.11: 1978
Requirements for detection devices. Rigid printed-circuit wiring.

BS 4737: Part 3: Section 3.12: 1978
Specifications for components. Beam interruption detectors.

BS 4737: Part 3: Section 3.13: 1978
Requirements for detection devices. Capacitive proximity detectors.

BS 4737: Part 3: Section 3.14 1986
Specifications for components. Specification for deliberately operated devices.

BS 4737: Part 3: Section 3.30: 1986
Specifications for components. Specification for PVC insulated cables for interconnecting wiring.

BS 4737: Part 4: Section 4.1: 1987
Codes of practice. Code of practice for planning and installation.

BS 4737: Part 4: Section 4.2: 1986
Codes of practice. Code of practice for maintenance and records.

BS 4737: Part 4: Section 4.3: 1988
Codes of practice. Code of practice for exterior alarm systems.

BS 4737: Part 5: Section 5.2: 1988
Terms and symbols. Recommendations for symbols for diagrams.

BS 5979: 1993
Code of practice for remote centres for alarm systems.

BS 6320: 1992
Specification for modems for connection to public switched telephone networks and speech-band private circuits run by certain public telecommunications operators.

BS 6707: 1986
Specification for intruder alarm systems for consumer installation.

BS 6799: 1986
Code of practice for wire-free intruder alarm systems.

BS 6800: 1986
Specification for home and personal security devices.

BS 7042: 1988
Specification for high security intruder alarm system in buildings.

NACOSS Codes of Practice

REG (Issue 2): NACOSS regulations.

NACP 0: Criteria for recognition.

NACP 1: Code for security screening of personnel. Now replaced by BS 7858.

NACP 2: Code for customer communications.

NACP 3 (Issue 2): Code for management of subcontracting.

NACP 4 (Issue 2): Code on compilation of control manual.

NACP 5: Code for management of customer complaints.

NACP 10 (Issue 2): Code for management of false alarms.

NACP 11: Supplementary code for the planning, installation and maintenance of intruder alarms.

NACP 12: Code for wire-free interconnections within intruder alarms.

NACP 13: Code for intruder alarms for high security premises.

NACP 14: Code for intruder alarm systems signalling to alarm receiving centres.

Underwriters Laboratories (UL). Essential Standards

UL 365: Burglar alarm units – police connected.

UL 464: Audible signal appliances.

UL 603: Power supplies for use with burglar alarm systems.

UL 609: Burglar alarm systems – local.

UL 611: Burglar alarm system units – central station.

UL 634: Standard for connectors and switches for use with burglar alarm systems.

UL 636: Hold-up alarm units and systems.

UL 639: Intrusion detection units.

UL 681: Installation and classification of mercantile and bank burglar alarm systems.

UL 827: Central station for watchman, fire alarm and supervisory services.

UL 1023: Household burglar alarm system units.

UL 1034: Burglary resistant electronic locking mechanism.

UL 1037: Antitheft alarms and devices.

UL 1076: Alarm system units – proprietary burglar alarm systems.

UL 1610: Central station burglar alarm units.

UL 1635: Digital burglar alarm communicator system units.

UL 1638: Visual signalling appliances.

UL 1641: Installation and classification of residential burglar alarm systems.

Other sources of information

In addition the reader can refer to other organizations that provide information and Codes of Practice, including:

- American National Standards Institute (ANSI)
- National Burglar and Fire Alarm Association (NBFAA)
- American Society for Testing Materials (ASTM)
- National Institute of Justice (NIJ)
- Security Industry Association (SIA)

2 Lighting

This chapter is devoted to the most commonly practised methods of employing lighting in the security and health care sectors. It ranges from a means of protection from intrusion by security lighting to safeguarding individuals from accidents or associated hazards in the dark by providing lighting. Lighting must also of necessity be employed in emergency directional evacuation, since when the normal lighting is lost a failsafe light-emitting directional trail must be provided to lead the way to safety. Equally automatic light control systems are invaluable for lighting areas which are entered by a person when switching the lights on (or off) manually is not required. These methods also come under scrutiny in this chapter.

2.1 Security lighting

Alarms, as dealt with in the previous chapter, are a tremendous deterrent and make life extremely difficult for intruders. It should be recognized that the objective is to stop an attempted break-in rather than to try and counter it at a later stage. To this end security lighting is a great aid.

Light is one of the best deterrents of crime and can be used to great effect for perimeter or corridor protection when employed auto-matically. Reliable low cost automatic light control systems may be utilized to detect an intruder entering an invisible passive infra red pattern formed by an unobtrusive sensor head. Once the sensor head has detected the intruder, it will trigger a controller or similar device to switch on a flood of light automatically. The system will latch and the lights remain on for a preset delay even after all movement has ceased. The security lighting can then re-arm in preparation for any further attempt to enter a prohibited area.

A basic automatic security lighting system will consist of a detector or sensor, electronic lighting control unit, daylight (photocell) sensor and perhaps a warning buzzer. However there are many self-contained units on the market which are easily installed, inexpensive and can be invaluable in simple roles (Figure 2.1).

Let us look at these self-contained devices first. Low cost lights were principally devised as a courtesy, convenience and safety aid and can be easily used to provide illumination in dark corners such as

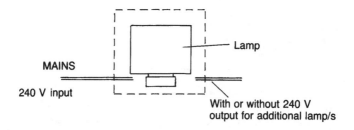

Figure 2.1   *Self-contained single unit construction*

garage and porch entrances. These automatic lights of single unit construction comprise of a lamp (very often globe lights or standard or mini flood lights), PIR detector and a daylight sensor which assesses the level of the daylight at the time. They provide a remotely switched automatic light only when needed, as once a moving body is recognized by the PIR detector the lamp will energize and provide adequate illumination for one to move around safely. Once the moving body is clear of the detection area the light will switch off after a prescribed period of time. These integral units generally have a 10 m–25 m PIR detection range depending upon the angle protection pattern they feature. Indeed most of these patterns are similar to those of their internal intruder detector PIR counterparts (Figure 1.4). However, there is also a semi-elliptical pattern that has a widespread use and this is separately identified in Figure 2.2.

These integral units feature a 24-hour or night only operation and

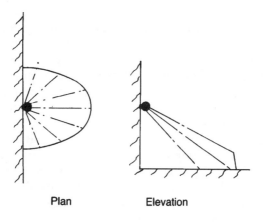

Plan   Elevation

Figure 2.2   *Semi-elliptical*

are wired to the normal domestic ac 50 Hz single phase supply. This should be through a switched fused spur unit originating inside the house using the appropriate fuse for the lamp load. Originally many units of this nature tended to have low output lamps and were intended for use mainly as a domestic device in low risk security applications. However as an extension of this type of unit, purpose designed integral automatic security lights are available. These have adjustable switch-off periods, various PIR detection patterns and are not only weatherproof but are also vandal resistant. Being a self-contained product, i.e. all components – lamp, PIR detector and daylight sensor – being formed into one unit, they are very easily installed. Normally the only adjustment required is that of turning the sensor head, which is mounted below the lamp head, to the correct angle and inclination for the required detection pattern. Integral automatic security lights often use halogen floodlights in order to throw a clear bright light and have a dual element PIR sensor in order to detect both heat and movement from either a body or a vehicle. Options on switching capacity can sometimes be provided for connecting further luminaires or floodlights via a relay output. Equally, voltage-free contacts can be used for switching devices such as closed circuit television (CCTV) and warning buzzers (Figure 2.3). The value of the single unit construction security light cannot be disputed as they are not expensive to buy, are easy to install and have variant detection patterns with adjustable sensors for ease of setting. Also many lamp types are available and as stated earlier further lights or other devices can often be wired to the master unit. These units are ideal for many applications but in some cases it is better to purchase a separate outdoor PIR light control device sensor and specific luminaires to suit the application (Figure 2.4). For example in some applications the sensor of a single unit construction device may not be in the optimum position when the lamp is. It may be difficult to adjust the sensor to view the required area since its position has been

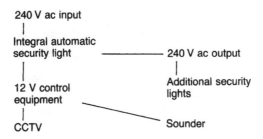

Figure 2.3 *Security light switching options*

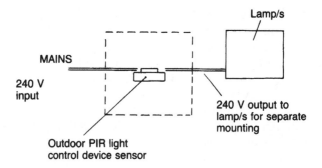

MAINS

240 V
input

Lamp/s

240 V output to
lamp/s for separate
mounting

Outdoor PIR light
control device sensor

Figure 2.4   *Separate outdoor PIR light control device sensor and lamp(s)*

governed by the position of the light, the latter's position obviously
being important. If the installer opts for a system with a separate light
sensor and luminaire, then the light or lights should be selected to suit
the particular area to be illuminated and the rated load of the light
control device. In certain cases filament lamps, globes or bulkhead
fittings could be used and in other cases halogen tube types could be
utilized. A suitable location for the light control should then be
determined based on its coverage angle which can be up to 270°.
These devices can also have any sensitive zone blanked if they
monitor an area which should not be viewed, such as an adjacent
neighbour's drive. The wires/cables should be routed from the lamp
or lamps to the terminal block via the gasket hole. The wiring method
is shown in Figure 2.5 and if required a manual override can be added
or indeed an additional detector can be incorporated. The terminal
block will be identified L.N.LS. respectively, live and neutral from the
power source, and live switched to the lamp load.

The sensor will have a time adjustment which controls how long

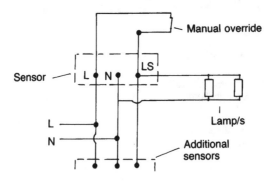

Manual override

Sensor

LS

L   N

L

N

Lamp/s

Additional
sensors

Figure 2.5   *Wiring 240 V PIR sensor*

the lamp will stay on after motion has been detected. The minimum time being the test time in order to perform a walk test in daylight conditions and to confirm correct wiring and function of the lamp. A light adjustment is normally also provided so that the desired lux status at which the control will shut off automatically can be selected by the installer.

The reader will now have an understanding of mains wired single unit construction security lights and also mains wired separate outdoor PIR light control devices with separate lamps. However there are also on the market light control systems which use a mix of 12 V wiring. These are often used where installing certain parts of the wiring to some of the sensors or ancillary equipment would be difficult with mains cabling, so low voltage parts are used and wired via signal cable as found in alarm or communications practices. Such a light control system can consist of two main sections which are secured together. First there is the control or junction box which is used to mount and connect the unit. It will contain the daylight sensor, control timer circuitry power relay and the terminals used in the wiring. The other main section is the sensor head which can be moved on its adjustable junction and is used to house the PIR sensor which operates from the 12 V inbuilt source. With these scanning and automatic switching device systems the power relay is brought into operation by an electronic timer circuit which determines the interval for which the lights stay on after detection of movement.

In view of the fact that this system resets at every detection, the connected lights will stay on as long as the person remains within the field of view. This time interval is adjustable but a walk test facility is normally provided to reduce the time interval to a few seconds. This allows expedient checking of the field of view during installation so that adjustments may be made.

Daylight sensors are always incorporated so that lights are only operated during hours of darkness but an option is usually provided so that they may also be operated during daylight conditions if required. Naturally a manual override should be provided so that the lights can be operated from a remote switching point. The wiring of this must very much depend upon the system that is being employed but an example of automatic operation with manual override is shown in Figures 2.6 and 2.7 employing Live, Neutral and Load terminals. Installation is straightforward; mains power is supplied from a fused source and taken via a conventional wall switch to the unit. The switch is an important part of the system as it provides control of the functional operation, the supply cable then being run to the lamps. One should appreciate, however, that the supply should remain in an active state hence it should not be switched in order to control the load.

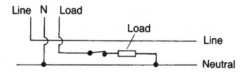

Figure 2.6 *Manual override off plus automatic*

With these automatic switching device systems the wiring is taken up to the control box but remote switching can also be employed. A variant on this system is that used in residential properties. This sort of system is purchased with a controller and a separate sensor head (Figure 2.8). The controller, which comes complete with an integral transformer, is mounted indoors and connects to the main power supply in place of the light switch. It is intended to blend into the domestic environment and is hence compact in form. The transformer supplies regulated 12 V dc power to the PIR sensors whilst the controller itself monitors their condition. The user control is a light switch 'on', 'off' or 'automatic' positions. There is also an adjustable 'on time' switch to select the desired lights delay once the controller has been triggered. With the light switch in the on position the lights will be illuminated at all material times and with it in the off position the lights will be extinguished at all material times. However with the switch set in the automatic mode the lights will come on during the hours of darkness when the sensor or sensors have been triggered and will remain illuminated for the period set by the adjustable 'on time' switch. An LED can be provided on the controller to show the state of the installed lights.

The sensor heads are intended for installation outdoors but in sheltered locations. Full coverage adjustment is allowed and a number of sensors may be connected to one controller. Only one sensor needs to be tripped for the light to switch on. A photocell in the sensor head automatically cancels any daytime operation but this is often optional at installation in the event that the user wants the system to function in a 24-hour mode. Other devices such as bell pushes or low voltage switches can be used to trigger the lights. These

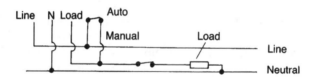

Figure 2.7 *Manual override on off plus automatic*

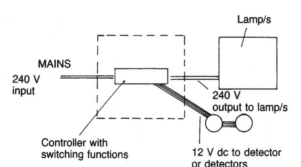

Figure 2.8 *Light control system with 12 V dc detectors*

may be normally open or closed depending upon the system. Such pushes allow the user to energize the lights without having to switch the controller from its automatic setting. In practice the bell push presents a short circuit, usually across the OV (zero volts) and trigger terminals to manually simulate the operation of the PIR sensor although it may be necessary to install a resistor of a recommended value in the wiring.

On leaving the premises the homeowner can operate the bell push in order to activate the lights, hence they are immediate and one does not need to wait for the sensors' operation for illumination to occur. This is an obvious advantage since the lights are on before the premises are vacated and the user does not need to take a number of strides in the monitored area for them to operate. The countdown cycle time will not commence until the protected area is finally vacated.

Many controllers can be programmed to switch on the lights automatically at nightfall and off in the morning by the simple use of a photocell, although inexpensive self-contained low wattage luminaires are also available for this specific duty, see Section 5.1. In order for the reader to appreciate some of the options provided by a typical system that features a sensor or sensors plus a controller with manual slide or rocker switches a wiring diagram is shown in Figure 2.9. This assumes that the controller has an inbuilt manual override switch for permanent operation of the lights. It may be found from the diagram that both Live and Neutral wires must be available in the existing switch box if one is attempting to retro fit the controller in place of an existing switch. Often it is not the case that both of these wires are available hence a separate mains cable must be taken to the unit. Having appreciated the application and use of the controller we now need to look at the sensor in a little more detail.

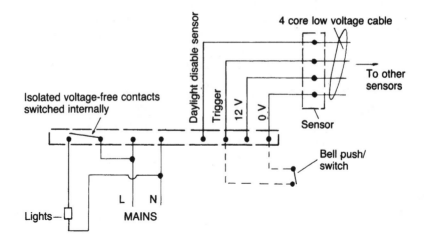

Figure 2.9 *Controller to 12 V dc detectors wiring*

2.2 Sensors

The selection and position of the sensor head is of paramount importance and the installer must take into account the detection pattern provided by the device if no troublesome blind spots are to be inherent in the installation. Obviously the lighting sensor can usually be seen by a potential intruder as it is exposed, unlike the sensor which is part of an internal alarm system and which can be placed in a hidden corner. For this reason much care must be exercised in its selection, especially for high risk areas. Certain sensors feature a rounded face so even if an intruder were to have knowledge of sensor types he would still find it difficult to predict the protected pattern, i.e. in which direction the PIR element was facing. The positioning is of even more importance for externally mounted sensors if access can be gained from above or behind in an attempt to disable them. In other cases physical restraints, such as barriers, may be needed to protect the device. The PIR detector is a commonly used sensor in security lighting as it is reliable, inexpensive and easy to set providing attention is given to the fact that its absolute range is subject to variation because of differences in background characteristics, type of clothing and ambient temperatures.

Recognizing that light control devices must be carefully situated for accurate operation and that they are not normally recommended for use in conjunction with audible alarm systems, because false tripping would be unacceptable in such circumstances, let us look at the guidelines for sensor location.

Hot and cold objects should never be in the field of view. Automatic sprinkler systems, heater flues or heat exchangers that are associated with air conditioner systems are clear examples. Strong sources of heat rising can trouble PIR detectors, equally the detector should not be placed directly above or in close proximity to the light source. The load and in particular incandescent lamps should be out of the PIR's protection pattern as the sensor may detect the lamps' change in temperature. A reduction in sensitivity will always be caused by any heating effect on the sensor although it is acceptable that light from the luminaire can fall on the daylight sensor since it will not cause the system to function incorrectly. Sunlight falling directly on the face of the sensor can cause permanent damage to the device and this can occur whether or not power is being applied by the system.

The sensor must be firmly mounted in position and not secured to any surface that could vibrate or move due to the wind. Sensor heads themselves can be locked, once adjusted, to ensure that they do not move on their base plate but this mounting must itself be extremely secure. Bushes and trees can cause turbulence during windy conditions and this can be a problem if they normally hide potentially warm areas or surfaces – even fences can fall into this category. The detector 'sees' a rapidly changing heat source moving in location and this can lead to sporadic operation.

Other obvious hazards are animals, especially if it is possible for them to approach close to the sensor although small birds and moths are unlikely to cause problems. Indeed it is unfortunate that the technology of passive infra red detection may never be capable of reproducible stable operation in outside applications because, although it is not the only detector in the security lighting field, it is probably the most popular. In fact its limitations can be compounded during daylight if 24-hour operation is required. A major problem is caused by the rising sun since there is no glass, as in internal applications, to filter out almost all of the infra red radiation from the sun. Also swinging signs or swaying trees will always cause nuisance false trips.

Having now obtained an insight into how security lighting is applied and how PIR sensors are best sited for optimum performance Figure 2.10 can show something of a cumulative result for an independent light control system. In this application the two areas that require protection are the front and side entrance doors which are best protected by volumetric and corridor detectors as shown. This figure shows how the controller, manual override switch and bell push are best placed adjacent to the front door and to how all the

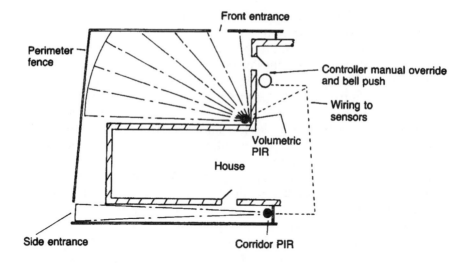

Figure 2.10   *Independent light control system application*

wiring, including that from the sensors, should be routed. The wiring is then taken from this location to appropriate lights and ancillary equipment if required. The manual override switch enables the lighting to be permanently energized. The bell push allows the authorized user to energize the lights when vacating the premises and to start the 'lights on timer'. Using this method the lights have already illuminated the protected area before the user leaves the premises and he/she does not need to walk even a short distance into the darkness.

Although this application shows the use of the popular PIR detector, other sensor types are also most worthy of mention. Doppler effect microwave (MWD) detectors employ special signal processing that enable many PIR false alarm causes to be avoided but careful siting is vital since microwave energy is not confined to the area being observed and can pass through many building materials. Also microwave detection is affected by external sourced signals such as radio and radar transmitters. Sensors that combine both MWD and PIR detection can be employed in areas where false tripping must be virtually eliminated. With these devices both elements must respond, hence motion and a rapid change in temperature must occur simultaneously. Alternatively a 12 V outdoor PIR can always be wired in parallel with any other type of detector including infra red (IR) beams to confirm a signal without any loss of detection ability or the compromising of security. IR beams are 'active' in the sense that a transmission of infra red energy is made by the detector system rather than just monitored in a given area as with the 'passive' detector. Infra red beam interruption systems come in two main parts – a transmitter and a receiver – and work on the principle that they are placed at a given distance apart so that any interruption of the beam projected between the two causes a response and activates the control.

These weatherproof units tend to feature two separate biaxial beams in order to minimize any false trips, otherwise falling leaves and wildlife in the vicinity could cause problems. The beams are set sufficiently far apart so that only a human size object could cause the required double interruption. Such IR beams usually have a range of some 100 m and do not need to be interconnected, i.e. they can be powered from separate 12 V dc sources and are wired so that the system cannot possibly be set unless both beams are clear. Using a two relay system a fault signal is provided for a blocked signal beam. Although these devices have a high immunity to sunlight, headlights and artificial lighting, it is still good practice to mount the unit where it avoids sunlight or artificial light falling on the optics as far as possible.

When the unit is fitted on a sloping surface the receiver should always be positioned at the higher point. Figure 2.11 gives examples of mounting biaxial beams. Housings are also available to accommodate multiple IR beam transmitters and receivers to form detection barriers with a covert array of beams. Alternatively units may be installed some 100 m or so apart and by linking them together one can achieve a cost effective detector system to cover many kilometres of perimeter protection.

The beams themselves are tuned for optical alignment using a voltmeter with the specific value dependent on the distance apart that the detectors have been installed. The lenses are kept clear of condensation by integral heaters as condensation could give rise to a weak and distorted signal.

Purpose built IR beams are also available for the exacting duty of working in hazardous areas where flammable gases may become present and which are therefore subject to special requirements. These units feature explosion proof and weatherproof housings which are carefully designed to prevent the beams' electrical circuitry from causing any spark that could ignite the surrounding flammable atmosphere. These devices are designed and manufactured in accordance with governing certification procedures and are often found in chemical processing plants and gas distribution centre applications.

In addition to convenience and security, another advantage of automatic security lighting is that it is also energy saving, since everyone at some time or another forgets to switch off lights when they are not required. In fact in the early days of these lights being

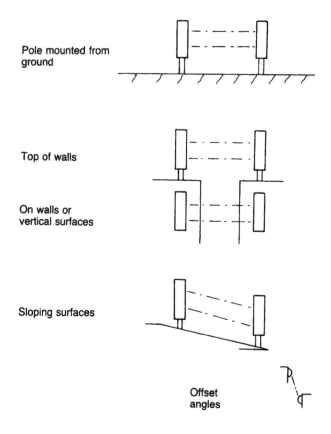

Pole mounted from
ground

Top of walls

On walls or
vertical surfaces

Sloping surfaces

Offset
angles

Figure 2.11   *IR beams*

available, it was said that few security lighting units were security
devices in the real sense but courtesy devices that also featured an
element of security.

   If one does not want to go to the expense and trouble of having an
intruder alarm installed then a basic security light control system can
be beneficial, even if only to protect the most vulnerable parts of a
house. For instance a sensor can be mounted to protect a patio door,
which is the point of a house often seen as the first forced entry target
for a burglar. Although modern domestic automatic light control
systems provide a quite adequate, usually 5 A, output, even heavier
loads can be controlled by energizing an external relay or contactor.
For commercial and industrial applications the lighting system can be
extended to great effect by driving audible signalling equipment
(with limitations), camera monitoring units and remote signalling
dialling equipment

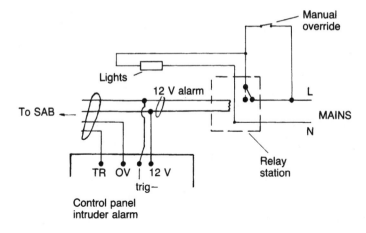

Figure 2.12   *Intruder alarm signalling lights*

Although audible outputs should only be connected to light control systems with discretion lights can often be switched on when an intruder alarm is signalled by an internally mounted detection device. This is illustrated in Figure 2.12 using a 12 V dc coil/240V ac relay. The system should not work in reverse, i.e. the light control system should not activate the audible alarm unless it forms part of a dual multi-detection system, which requires a secondary activation to occur before the alarm condition is generated. Even then it could only be employed in an area in which no authorized user would be prior to the alarm being switched off.

Figure 2.13 shows how a latching personal (PA) push can be used to provide an audible and visual output by using a 12 V power supply

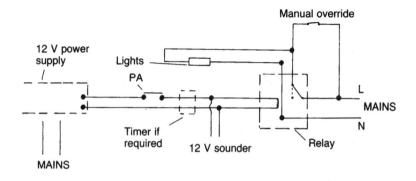

Figure 2.13   *PA signalling audible and visual output*

and relay that switches the Live until manually reset at the push. Alternatively an inexpensive and readily available timer module that provides a switched output for an adjustable period of time once triggered by a 12 V pulse could be interfaced. This will turn off the output to the sounder and relay after the selected time has expired. An override is shown in Figure 2.13 to enable the lights to be on for normal use. This inexpensive method can be used to protect cash tills in small commercial premises by providing an immediate audible and visual alarm if threatened. If an intruder alarm control panel is already in existence then the power supply is not needed and the PA push is simply wired into the appropriate PA output terminals and is subject to the timer reset provided by the control panel. The 12 V coil/240 V ac switching relay is then wired into the control panel's bell/sounder output terminals. The 12 V sounder is also wired into those terminals and the lights have the Live switched through the main relay contacts which go closed in alarm. The neutral terminals are tied to the same point. If the power supply method is used then the PA contacts must be capable of handling the sounder load, hence a microswitch type PA is needed and not a reed contact version.

2.3 Occupancy detectors

A highly viable simple switch which can also act as a security deterrent to intruders is the occupancy detector. These are gaining in popularity and are extremely easy to install.

Although intended for energy saving by ensuring that lights are not left on in unoccupied areas, occupancy detectors are also useful for reasons of convenience, security and safety by providing automatic lighting when an area is entered by a person. They are often found at the entrance to corridors or store rooms where manual switching is not required. Equally they can be utilized at the entrances to unattended areas where illumination can act as a security deterrent to intruders.

The occupancy detector which is intended only to turn lights on when necessary, uses the PIR principle to operate an electrical relay switch. It includes a time lag to enable continuous operation to be affected for a period often adjustable from some 10 seconds to 10 minutes. A photocell, itself adjustable, prevents the relay operating when sufficient light is available. The relay, although intended in the main for connection to lighting loads, can control other electrical loads. There are three occupancy detector switch types: the PIR wall switch, PIR occupancy switch or PIR ceiling switch. These are often referred to as automatic security switches.

PIR wall switch

This is intended to replace an existing wall switch by detecting persons passing in front of it. It features a horizontal detection range and is ideally suited for positioning at the entrance to an area where 'no hands' light switching is required.

PIR occupancy switch

The PIR occupancy switch or space switch plugs into a ceiling socket and provides a detection pattern normally forming a 90° arc with a range of some 10–15 m. Its method of installation is similar to that of the PIR ceiling switch.

PIR ceiling switch

A PIR ceiling switch is mounted above doors and in corridors and is installed next to a luminaire using a special socket. This socket is wired to the existing terminal block so that the existing switch and wiring continues to be used. It recognizes a person passing underneath the detector. Once installed next to the luminaire using a special socket a plug-in receiver switch is connected to provide the required switching. The appropriate socket should be wired to the terminal block within the first light fitting of any group as in Figure 2.14.

2.4 Lighting types

Accepting the importance of light as a first line of defence against all forms of crime, permanently energized security lighting can be seen

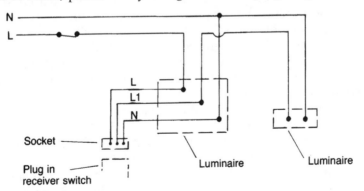

Figure 2.14 *PIR ceiling switch circuit*

in a dual role. An intruder risks detection whether he leaves it on or attempts to turn it off. However the exact level of illumination to be provided very much depends on the area and nature of the premises to be protected. Many lighting types are available and although high output levels are usually employed for timed automatic security lighting, low energy lighting is used when a system is intended to be energized permanently during the hours of darkness.

Almost any type of light can be used for security purposes although tungsten halogen floodlights are often found because of their good illumination properties and because they light immediately. The most popular form have either 150 W, 300 W or 500 W bulbs and are often found in die cast aluminium housings that are enclosed by a glass front, a metal front grid, or have the front open to the atmosphere. These are for wall, floor or pole mounting and generally have a quick release front facility to give easy access for lamp replacement. Higher output outdoor floodlight projectors can accept 750 W, 1000 W or 1500 W tungsten halogen bulbs and these can be found in more specialist applications.

Tungsten halogen floodlights and projectors are very popular in compounds and storage areas. One disadvantage is that they have a short lamp life which can be disastrous where they are essential for security, unless they are supported by other lights. Also they are expensive to run over an extended period. However, for vulnerable areas, their good illumination properties can be used to great effect and in the domestic sector mini types are often mounted at first floor height on houses. Unfortunately they are not always practical on bungalows because the lower the mounting height the greater the glare. Looking at 'glare' in more detail it is recognized that the main objective of security lighting is to provide illumination in a given area so that a burglar or vandal can be easily detected by a neighbour, police officer or security guard. This glare can in fact be used to great effect by restricting the vision of intruders providing it does not impair the vision of security personnel. Of necessity the total area to be protected must be illuminated to a sufficient level but one must take care to ensure that interior lighting does not affect the view of someone who is monitoring the area from an inside vantage point. In this context the reflections from windows must be kept to a minimum. Also care must be taken to avoid shadows being cast especially in storage areas. As a general guide 300 W to 500 W tungsten halogen floodlights should be mounted at a height of between 3 to 5 m whilst the larger wattage units should be positioned higher.

In many domestic applications modern bulkhead lights are used. These are capable of controlling 100 W bulbs and they feature polcarbonate or die cast enclosures which are not expensive and provide quite adequate illumination levels. In addition to security lighting they are also good for amenity lighting, the bulb life is good and it can be easily renewed by the occupant of the premises. These units resist impact and abuse reasonably well and for low security applications their use can be recommended.

In areas where the lighting must run for long periods low energy lighting must be used with its inherent long lamp life. Modern low energy, low maintenance discharge and fluorescent lamps provide extremely cost conscious yet effective security lighting since filament lighting is expensive to run over an extensive period and the short lamp life is a problem.

Three of the modern lamp types that are particularly suitable for security lighting are high pressure sodium, low pressure sodium and compact fluorescent. High pressure sodium which has an efficiency of some 120 lumens per watt is now a dominant light source. It provides an extremely efficient golden white light source which allows colours to be distinguished and it is often found in CCTV surveillance installations.

High pressure sodium can also be used for security and amenity lighting where long burning hours and good colour rendering are needed. Factory buildings, car parks, gatehouses, access roads and general amenity areas come into this sector. Lanterns that are corrosion resistant, weatherproof, lightweight and vandal proof are ideal for such a purpose. If equipped with a photocell, automatic switching ensures that the light is only active at night.

As a general rule a 70 W lamp should be mounted at a height of between 3 and 5 m, whereas a 100 W, 150 W or 250 W unit should be mounted progressively higher. A 400 W unit requires a height of 10 m bearing in mind that anything any lower will significantly worsen glare.

Low pressure sodium has a distinctive monochromatic yellow light form and is matched to the maximum sensitivity of the eye to give optimum vision at low light levels or in poor visibility. It is recognized by many as the most efficient light source available and can be used to protect small areas at a reasonable cost.

Compact fluorescent is ideal for commercial indoor security lighting. It has a small light source which matches filament lighting in both colour and quality but has the advantage of consuming less energy and having a longer lamp life.

An excellent way to deter potential intruders is to use well-lit perimeter fencing. If it is not possible to use a manual perimeter patrol then lighting will always permit the use of CCTV (see Section 3.4). In cases where the objective is to detect rather than deter, cameras that are sensitive to infra red light may be used so that intruders are not able to see how extensive a security system is or in which location they are most vulnerable.

Another alternative is to protect the premises with a continuous zone of light that creates a barrier that cannot be cheated. If it is lit the intruder is risking detection. If he removes the light source, the darkness itself will warn security staff.

This zone of light technique can apply to domestic premises as well as commercial and industrial practices. When correctly applied, domestic security lighting not only helps neighbours detect intruders when the family are away from home but it enables people to move safely around the grounds.

Although external detection systems which signal and energize lighting only when triggered by the movement of a person or vehicle in a defined area remain common, the use of low energy lighting being run for extended periods is now becoming extremely popular, indeed its use is set to grow. We can see this extended period lighting taking over a lot of the roles traditionally employed by automatic lighting that brings lights on only when movement is detected.

On the basis of the growth of this extended period lighting we can summarize the characteristics of the principal forms. This lighting type may be controlled by separate or integral photocells and timers or by simple manual switching.

Extended period lighting

High pressure sodium (SON)

Classified alongside discharge lighting. It provides an extremely efficient golden white light source and is often found in CCTV applications as it allows colours to be distinguished. Ignitors are required to start up the light source. Ratings: 70, 100, 150, 250, 400 and 1000 W, up to 130 lumens per watt.

It is available as SON-E-elliptical which is a coated elliptical bulb to minimize glare as it is dispersive. SON-T-tubular should be used for floodlighting luminaries and super-critical photometric fixtures.

Twin-arc tubular has a double arc tube to guarantee immediate restrike after a power cut. It tends to be adopted in areas where maintenance is difficult or expensive. Superimposed triple pulse ignitors are required.

Low pressure sodium (SOX)

Classified alongside discharge lighting. It is a high efficiency monochromatic source with low running costs. The yellow light-form is matched to the maximum sensitivity of the eye to give optimum vision at low light levels or in poor visibility. It is often used to protect small areas at a reasonable cost. It is unsuitable for applications where colour discrimination is required. Ignitors are required to start up the light source. Ratings: 18–90 W, lumen output 1800–10 800.

Compact fluorescent

Ideal for commercial security lighting. Has a small light source which matches filament lighting in both colour and quality but is less energy consuming. It is low level security only and there is a huge variety of types and styles of luminaires.

Metal halide

Classified alongside discharge lighting. It offers a clear white light and is suitable in floodlights when the heat generation of tungsten halogen would create a problem. Ignitors are required to start up the light source. Ratings: 70, 150, 250 and 400 W, lumen output 5000–31 000.

It is available as HQ1–E which is an elliptical bulb giving a soft light distribution with less glare. HQ1-T is a clear tube mainly used in low bay lighting.

Mercury fluorescent

Classified alongside discharge lighting. It is an alternative to conventional fluorescent lighting, giving a cool white light for applications that need pleasing colour quality. Bulbs are elliptical with coated inside surfaces of europium yttrium vanadate phosphor. It must be operated with appropriate control gear. Ratings: 80, 125, 250, 400, 700 and 1000 W, lumen output 3700–58 000.

2.5 Environmental protection

The choice of components used for security lighting and especially those components needed for outside duty are, of necessity, governed by their ability to withstand extremes of climate, changing ambient temperature limits and by their resistance to the weather and ingress of liquids and dust. If they are to perform as required then they must be designed and mounted with these factors in mind. It is totally pointless to have very good levels of security lighting and then to discover that the system failed to operate because the equipment used did not have the correct level of environmental protection.

Installers should therefore select apparatus with the required degree of protection to the weather and to the ingress of liquids and dusts. If this is done wisely then the positioning of components can be more carefully exploited. International Standards detail degrees of Environmental Protection and these are recorded in Table 2.1.

The specific test method is detailed in the International Standard and this should be followed precisely in order to prove the apparatus protection level. For reference sake the degree of protection is indicated by the letters IP followed by two characteristic numerals. The first numeral indicates the protection afforded against the ingress of solid foreign bodies and the second the protection against the ingress of liquids. A common rating is IP54, which offers protection against the ingress of dust to the extent that it cannot enter the enclosure in a quantity that could interfere with the satisfactory operation of the equipment. Water splashed against the equipment from any direction will have no harmful effect.

It should be understood that many goods are supplied in housings which are coded to indicate their classification when gaskets are correctly seated and the conduit is sealed on wiring. It follows that in a pre-installed state they do not necessarily warrant the applied code.

On occasion one may also see supplementary letters applied to the IP code. The letter to note particularly is W which draws attention to the fact that the equipment so designated is suitable for use under specific weather conditions. The unit will have been provided with additional protective features or processes. The specified weather conditions and noted features or processes are to be agreed between the manufacturer and user and can be endorsed in data provided by the said manufacturer. For instance one could draw attention to the circuit board that may be additionally lacquered for water resistance.

In North America the National Electrical Manufacturers Association (NEMA) system is recognized for classifying enclosures for industrial controls and systems. The classifications are contained in Publication No. 1CS-6. This standard is used principally in the USA and adopted in the public interest to eliminate misunderstanding between purchasers and manufacturers. In view of the fact that many countries throughout the world do adopt North American classifications it is always wise to have a little understanding of their meaning. Particular attention must be paid to switches that can be used to control many types of system including lighting.

Type 1 general purpose indoor

These are intended for use indoors, primarily to prevent accidental contact of personnel with the enclosed mechanism in areas where

Table 2.1 *Degrees of environmental protection*

First characteristic numeral	Degree of protection: *solid bodies*	Second characteristic numeral	Degree of protection: *liquids*
0	No protection	0	No protection
1	Protection against ingress of solid bodies larger than 50 mm, e.g. accidental contact by hand	1	Protection against drops of condensation
		2	Protection against drops of liquid falling at an angle up to 15° from the vertical
2	Protection against ingress of medium size solid bodies larger than 12 mm, e.g. finger of hand		
		3	Protection against rain falling at an angle up to 60° from the vertical
3	Protection against ingress of solid bodies greater than 2.5 mm thick	4	Protection against splashing water from any direction
4	Protection against ingress of solid bodies greater than 1 mm thick	5	Protection against jets of water from any direction
5	Protection against harmful deposits of dust. Dust may not enter the enclosure in sufficient quantities to interfere with the satisfactory operation of the equipment	6	Protection against water similar to conditions on ships' decks
		7	Protected against the effects of immersion in water for specified pressure and time
6	Complete protection against ingress of dust	8	Protected against the prolonged effects of immersion in water to a specified pressure

Note: When it is required to indicate a class of protection by only one characteristic numeral, the omitted numeral shall be replaced by the letter X. For example IPX5 or IP2X.

unusual service conditions do not exist. Protection is provided against falling dirt. Enclosures may be non-ventilated or have ventilated openings.

Type 2 drip proof indoor

Intended for use indoors to protect the enclosed mechanism against falling non-corrosive fluids and falling dirt. These enclosures have provision for drainage. As with Type 1 they may be non-ventilated or ventilated.

Type 4 watertight and dust-tight indoor and outdoor

These are developed for use either indoors or outdoors to afford protection against splashing or hose directed water, seepage of water or severe external condensation. They are non-ventilated and sleet resistant but not iceproof. They do not afford protection to the accumulation of ice interfering with the successful operation of the equipment. This type corresponds generally with IP54 but in addition is corrosion resistant.

Type 6 submersible, watertight, dust-tight and sleet resistant indoor and outdoor

These are intended for use indoors and outdoors where occasional submersion is encountered and corresponds generally with IP67.

Type 13 oil-tight and dust-tight indoor

These are intended for use indoors and must afford protection against lint and dust, seepage, external condensation and spraying of water, oil and coolant. This type corresponds generally with IP65.

NEMA defines 'non-ventilated' as a means so constructed as to provide no intentional circulation of external air through the enclosure. Ventilated enclosures normally have the same provision as non-ventilated types except that they may have a ventilated opening, which must however be able to prevent the entrance of a rod of a specified diameter.

By quoting IP or NEMA classifications, ambiguous misleading terms may be avoided and installers can ensure that no attempt can be made to position unsuitable components in troublesome areas. BS 5420 'Degrees of Protection of Enclosures for Low Voltage Switchgear

Drip proof   Splash proof   Watertight

Figure 2.15 *Ingress protection marking*

and Controlgear' contains criteria on IP classifications and can be invoked for many components used in the security industry and intended for outside duty.

Very often the various components used in lighting systems, especially intended for outside use, do have an IP marking applied to them either by means of indelible ink, by labelling or the classification is moulded into the body. More general symbols can at times be found, often on goods imported from other parts of Europe. These marks are shown in Figure 2.15.

Security lighting offers the additional benefit of safety provided that it is correctly specified and installed with due regard given to all conditions. However emergency lighting as detailed in the following section is intended only for safety purposes to ensure that illumination can always be available in prescribed areas.

2.6 Emergency lighting

One has only to experience an electricity power cut at home to realize the problems that can occur when no mains supply is available. This problem is compounded during the hours of darkness and even more so when the power cut has not been forecast and persons are caught unawares. Such a situation is not only undesirable but for obvious reasons it is highly dangerous. The reader will appreciate how much greater the problem is for establishments that are open to the public, to provide amenities or entertainment. The task of emergency lighting is to provide standby lighting by means of battery packs in the event of such a power failure or failure of the normal mains lighting. This failure can often be criminally induced by deliberate fusing of the lighting in public areas. Although emergency lighting is not intended to and does not provide illumination levels to the same degree as the normal lighting, it does however provide adequate light to fulfil its specific standby purpose for the building type.

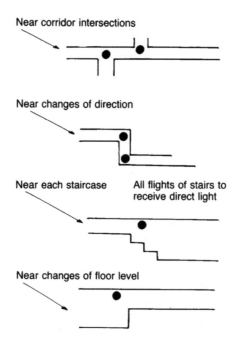

Figure 2.16   *Positioning of emergency luminaires*

Siting of luminaires

The correct positioning of emergency luminaires (Figure 2.16) is of paramount importance to ensure that the system provides a safe and effective means of lighting. Also the system of lighting must comply with all relevant legislation.

In the main luminaires should be so positioned in order to:

(a) illuminate exit routes clearly: an escape lighting luminaire or sign should be sited near each exit or emergency exit and outside each final exit point;
(b) illuminate intersections of corridors and changes of direction or floor levels which may constitute a hazard;
(c) illuminate all fire alarm call points and fire fighting equipment;
(d) illuminate all passenger lift cars: only self-contained systems are acceptable for this duty.

In addition luminaires must also be placed in toilets exceeding 8 m². Also in

- plant switching and control rooms;
- foyer and reception areas;
- kitchens or areas where certain processes could be halted on mains failure.

The emergency lighting system should be designed so that a minimum level of illuminance of 0.2 lux at floor level along the centre line of the escape route is achieved, but a higher level should be achieved if risks are present. Such a lighting level allows obstructions to be seen even though theoretically there should be no obstructions on escape routes. However in premises such as supermarkets or factories, obstructions could be present during the emergency period. Therefore, at an early stage, consideration must be given to the building type, its uses, and the necessary lighting levels. For instance old persons' homes or hospital geriatric wards may need levels above 0.2 lux, since residents will almost certainly be physically incapacitated. Large open plan buildings have an irregular furniture layout so the escape route can be difficult to define. Hotels are normally occupied by people on a short-term basis and guests may be totally unfamiliar with the layout of the building. In premises such as supermarkets, cinemas and theatres, additional lighting at cash desks, etc. is necessary for security purposes.

Before the installation of any emergency lighting system can begin the legal requirements must be considered because there is a huge volume of legislation covering the various types of premises. Many principal statutory documents do exist but in addition to national standards many local authorities publish their own recommendations. In all cases the Fire Prevention Officer should be consulted as he or she will have the enforcing authority.

It will be found that many different types of luminaire are available for emergency lighting installation. In practice these should all be manufactured in accordance with Industry Committee for Emergency Lighting (ICEL) Standards, which are the industry standards for this form of equipment and which are monitored by the British Standards Institution.

Circular form luminaires are the most popular type in use. They are efficient, inexpensive, give an effective light source and blend easily into their surroundings. Under normal circumstances they feature an injection moulded base and diffuser although the latter can be specified in crystal glass. High output krypton lamps are usually employed in circular luminaires. Tungsten lamps or fluorescent tubes are the elements often used in the variously sized, rectangular shaped luminaires in common manufacture. The selection of lamp type and size are determined by the light output needed for the protected area.

To these can be added a screen printed diffuser making generally recognized 'Exit', 'Emergency Exit' or 'Fire Exit' signs or the running man pictogram.

For external use, in settings such as final exit doors, external fire escapes, walkways, car parks and transport depots, versatile weatherproof bulkhead fluorescent tube types are advocated. They have protection against the elements, are vandal resistant, and feature flame retardant enclosures.

For internal industrial applications where a versatile source of emergency flood lighting is needed, portable lights can be used where the installation is to be of only a temporary nature. These devices are often of twin floodlight form, using 12 V tungsten projectors which are illuminated automatically on mains failure by means of a maintenance-free, sealed lead acid recombination battery. Built to meet ICEL Standards these units have a steel housing which encases the battery and control unit. By means of an integrated circuit regulator and protection circuit the battery is protected from the deep discharges that could occur. Integral LED's monitor mains, charging levels and the state of the lamp filaments. These units are invaluable for placing in narrow walkways that may require an emergency lighting level higher than that for a simple escape route, because of the obstructions that could exist in an industrial environment.

Emergency lighting is classified into one of two systems: 'self-contained' or 'central battery'. The latter is employed for the medium or larger installation.

The self-contained system uses luminaires that have all the required elements built into a single housing. This and the central battery form are detailed at a later stage in the chapter, however the reader must first appreciate the three forms of self-contained systems. These are namely non-maintained, maintained and sustained.

Non-maintained

In this type of system the luminaire which contains one or more lamps which are normally off will automatically operate from the emergency supply only upon failure of the normal mains supply (Figure 2.17).

Maintained

The luminaire, in this system type, which contains one or more lamps will operate from the normal ac supply or from the emergency supply at all material times.

Many maintained luminaires are supplied with a 4-way terminal block to enable both switched and unswitched incoming mains

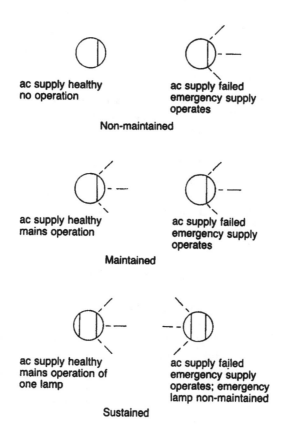

Figure 2.17   *Self-contained lighting types*

supplies to be accommodated. This means that the luminaire may be switched independently of the emergency lighting facility. The lamp which is normally on will automatically switch over to its standby power and remain on after any supply failure (Figure 2.17).

Sustained

The sustained emergency luminaire contains two or more lamps at least one of which is energized from the emergency supply and the remainder from the normal supply. Such luminaires are intended to sustain illumination at all material times. The emergency lamp is non-maintained only.

The emergency lamp is normally off with the independent lamp operating from the mains supply. In the event of supply failure, the emergency lamp will illuminate (Figure 2.17).

Self-contained systems

The self-contained unit features a lamp, a control unit including automatic charger and a standby element all contained within the luminaire housing. The standby element is usually a long life, high temperature nickel–cadmium (Ni–Cad) battery which is 'held off' electronically by the mains supply under normal circumstances. A charge monitor LED is incorporated to monitor the battery charging and lamp condition. For a wide variety of installations the self-contained luminaire is the most versatile form of emergency lighting. Since the battery that provides the power for illumination and is brought into operation automatically during the mains failure is contained within the luminaire, it enables installation to be carried out simply and economically.

Advantages of the self-contained system are:

1 Ease and speed of installation. Extending the system is easy to effect.
2 All units work independently of each other.
3 Little maintenance is required with the exception of periodic functional checks. Installation is low cost and the operation of the wiring generally safe.
4 Protected wiring is not needed and there is great system integrity.

The disadvantages of the self-contained system are that battery life is restricted to a maximum of 8 years and the ambient temperature range of the luminaire is only from 0°C to 25°C. Also unnecessary discharge of standby batteries cannot be stopped during mains failure in daylight conditions.

The wiring is relatively straightforward and should generally be in accordance with IEE Wiring Regulations, and with requirements that could apply to the building type and local by-laws.

The supply for the self-contained system should be derived from the local light source and the cables employed should be similar to those used for the normal mains lighting. In the event of a fire, emergency lighting wiring would therefore fail safe, i.e. become active if the fire is sufficient to damage the main lighting wiring.

Self-contained emergency luminaires should therefore always use PVC sheath. Although the supply to self-contained luminaires should be such that any unauthorized disconnection is prevented, a suitable means should be employed for simulating a failure of the mains.

On the basis that the same local fuse is used for the emergency lighting as is used for the normal lighting (Figures 2.18 and 2.19) the emergency lighting will be brought into operation if the normal

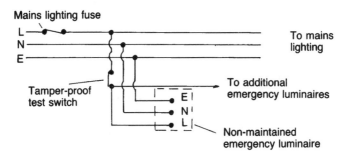

Mains lighting fuse

L — To mains
N — lighting
E —

Tamper-proof
test switch

To additional
emergency luminaires

E
N
L

Non-maintained
emergency luminaire

Figure 2.18   *Non-maintained system*

lighting fails in a given area. The connection method of the remote emergency housing to the mains supply is most important and one of the circuits shown in Figures 2.18 and 2.19 must be employed for test purposes.

Following installation, reference should be made to BS 5266 Part 1 'Emergency Lighting. Code of Practice for the Emergency Lighting of Premises other than Cinemas and Certain other Specified Premises used for Entertainment' as certain tests to ensure correct functioning should be carried out. In the Standard, recommendations for the clear indication and safe level of illumination of escape routes in the event of failure of the normal lighting are entered, plus recommendations are given for the minimum continuous periods of operation based on size, type and usage of the premises. In addition new common European designating orders are being developed which will strengthen legislation. The operation of each luminaire should be tested twice a year using the standby supply for at least 1 hour. Also once a month all luminaires should be energized for a short period from the standby battery to ensure correct operation.

Every 3 years the system should be energized for its full emergency duration. This is the manufacturer's declared duration for a battery operated emergency lighting unit and is the time that it will continue to operate after mains failure. This can be for any reasonable period but standard times are 1 to 3 hours.

The duration required by legislation is based on whether the building will be evacuated or not if just the mains fail. Whilst it is possible to use a one hour system if evacuation is immediate, it also means that the building cannot be re-occupied until a full recharge of the standby batteries has occurred. In practice It is normally

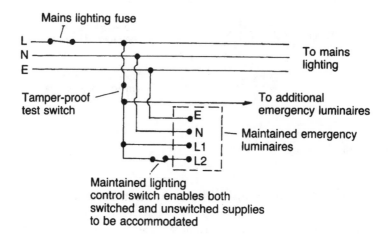

Figure 2.19   *Maintained or sustained system*

preferable for the installer to select a luminaire rated to operate for 3 hours after mains failure.

Of necessity the minimum escape route lighting levels must be achieved at all times and not just when the luminaire is new and the battery fully charged. Therefore attention must be given at all times while testing the system to the minimum illuminance requirements laid down in BS 5266. Scheme planning spacing/height data are therefore needed. Tables are provided by manufacturers to give nominal emergency lighting lumens over given time periods for different luminaire types. This data should be provided by a BSI registered photometric laboratory to give assurance that the designed performance will be achieved in practice. ICEL 1001 is the Industry Standard for photometric performance and claimed data of emergency lighting as tested by the British Standards Institution. Spacing tables provide the information to determine whether fittings are needed in addition to those required for the points of emphasis. In practical terms lightmeters are available for continuous field use and enable the operator to measure the luminance of emergency lighting systems across the range defined in the British Standard. The only things to consider when using such instruments are that the light output level measured should be two or three times the minimum to ensure that the standard will be met over the normal working life of the system. The impact of deterioration in performance, which will surely occur until the system is next tested, must also be accommodated.

As stated earlier the self-contained luminaire is used to great effect for small up to the medium application but the central battery system must be used for larger applications.

Central battery systems

Installations of a medium to large size, especially when numerous lighting units and signs are required, often necessitates a central battery system. This system has certain advantages including being more easy to test, having a comparatively low cost per unit of power on large systems and having an extended system life. Routine maintenance is easily achieved and luminaires may be sited in more diverse ambient temperatures. However central battery installations do have inherent disadvantages. For instance these systems need essential maintenance, are not easily extended and the distribution cables must have special protection or feature specific sheathing. Sub-circuit monitoring is necessary and cables are subject to selection in order to avoid voltage drop.

A central battery unit is a source of emergency power with its own battery, charger and mains failure contactor housed in one or more cubicles. The five major types of battery in current use are discussed below.

Sealed lead acid recombination

These are compact, fully sealed and provide an economical solution where a planned routine maintenance method is not apparent. They are often found in environments where gas discharges during charging cycles are not acceptable.

Tubular plate

Also compact but can be used where regular testing is a necessity. Such batteries permit delivery of high output currents over short time scales. They have a life of 15 years.

Flat plate

Of lead acid construction, these are suitable where long standby periods between discharges can occur. They exist as single cells and tend to have transparent cases with electrolyte level indicators. They are suitable for long spells of unaltered operation and have a battery life of up to 12 years.

Planté

These are of lead acid construction, with a long life and low maintenance capability and can give a life of up to 25 years, providing either a long or short discharge duty.

Nickel–cadmium

High suitability and particularly robust. Can satisfy diverse temperatures and need only simple routine maintenance. Unlike lead acid versions they can be stored or left in a discharged state without any damage occurring.

Luminaires for use with central battery systems are identical in construction to the self-contained luminaires described in the earlier part of this chapter. They are designed to operate continuously from the central battery system supply. By the use of purpose built conversion modules standard fluorescent luminaires can be converted to emergency luminaires for connection to the central battery system. Linked to a suitable central battery unit the module provides an efficient emergency lighting system using normal light system luminaires.

2.7 Monitoring

Manufacturers of emergency lighting equipment often provide charts that can detail such conditions as life expectancy, maintenance intervals and resistance to abuse. Volumetric sizes and weights also shown in such a form allow the designer or installer to select an appropriate battery system.

When these values are equated against charts showing emergency durations for available voltages and wattages related to the various battery construction types the system design may be completed. For non-maintained central systems the monitoring of lighting circuits (Figure 2.20) is effected by introducing relays, either in control boxes

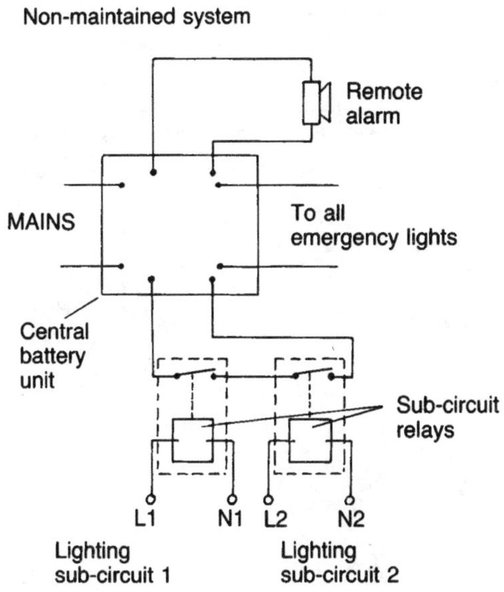

Non-maintained system

Figure 2.20 *Circuit failure brings on all emergency lights*

sited close to the distribution boards or by electrical relays in the central battery cubicle. Many circuits can thus be monitored in an individual way. By using this method any circuit failure will cause the total emergency lighting system to function. In maintained systems monitoring is not necessary, as in the event of sub-circuit failure lighting is already present. However it is possible that there are locations in a building where emergency lights are not desired. In such circumstances hold-off relays can be introduced; then if a sub-circuit fails but the mains supply at the central battery unit remains present, the relays energize the lights on that circuit from the low voltage from the transformer in the central battery unit. This battery continues to be charged and available in the event of total supply failure.

In the event of a sub-circuit failure the low voltage ac supply will control the emergency lighting whereas in the case of total power failure the battery supply provides the full power (Figure 2.21). Certainly it is an essential feature of any emergency lighting system that the lights will not only become active on mains failure but also on sub-circuit or fuse failure. This must be proved by a monitoring process and routine testing. Effective emergency lighting is an invaluable safety aid when kept in good order. Without doubt its application is

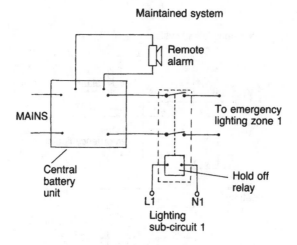

Figure 2.21 *Circuit failure brings on local emergency lights*

expanding to all new premises generally open to the public and to refurbished buildings. There is no reason why emergency lighting may not be utilized in the home even if only to the extent of installing one luminaire in an appropriate position which could prove hazardous if an unexpected cut in power was to occur.

An extension of emergency lighting is the light management system often now seen in commercial and industrial buildings. It works by automatically switching off lighting when the daylight that is present in the room increases the light level above a certain value. Once the light level falls below the determined value then the management system will once again switch on the artificial light sources. A microprocessor is used to evaluate a series of readings over a certain time period and is then programmed to act on a cumulative result before switching off the installed lighting. In view of the fact that this sort of system was initially designed to reduce running costs the light sensor will switch off the lighting if the daylight has been continuously below the designed light level for an excessive period. This ensures that they cannot be left on inadvertently during the night, although a manual switch can override the light management system control unit at any time. Hence the lighting management system can be designed with an element of safety and security whilst being cost saving in its application.

We mentioned earlier the use of alarm system infra red beams in hazardous areas, where special requirements exist for all electrical equipment in places where flammable gas or vapour can become present. These special requirements must of necessity extend to lighting for safety and security purposes.

The most effective way of illuminating any working area must be the obvious one-electric light, but what if a flammable gas were to become present? In that case the presence of electricity could be hazardous because any sparking which could occur (for example between a bulb and its holder) could cause an ignition leading to explosion or fire. Special purpose lamps feature a turbine which is driven by compressed air and coupled to a permanent-magnet alternator to power a mercury vapour bulb. These units are designed so that all of the electrical circuitry is contained within an enclosure which is filled with the exhaust air from the turbine, the electricity being made available only when this condition is achieved. Any explosive or flammable gases in the area of the lamp are thus prevented from reaching the electrical circuits. In view of the fact that the lamp is fed from an air line and not via any electrical cables, the source of power is not electrically hazardous. The application of these lamps is for inspection, maintenance and emergency lighting on offshore production platforms, ocean-going product carriers, petrochemical tank farms, gas distribution centres and chemical processing plants.

Further flameproof or explosion protected 240 V luminaires are also available using different light sources. For emergency lighting purposes these incorporate an integral emergency dc power supply which is recharged from the mains. These luminaires maintain lighting for escape routes during emergencies and are designed so as to withstand severe hostile environments. They must be chosen for the application with specific attention having been given to the zone itself, the probability of gas becoming present and the gas subgroup involved.

Before becoming involved with any type of lighting, not just that for hazardous areas, it is essential to recognize the requirements of the Health and Safety at Work Acts, although essential guidance for safe installation, maintenance and disposal of lamps and lighting equipment tends to be provided with quality equipment.

For the general requirements and tests of luminaires British Standard BS 4533 is the governing document. It contains requirements for the classification and marking of luminaires, for their mechanical and electrical construction and photometric performance together with related tests. It is accepted by HM Government as a 'safety' specification and the British Standards Institution (BSI) Safety Mark (Figure 2.22) gives an independent guarantee to all concerned that the luminaire has been designed and made in accordance with good engineering practice. Equally it has been type tested and its

Figure 2.22   *Safety mark*

manufacturing quality is monitored regularly by Inspectors of the Quality Assurance Department of the BSI.

The best possible guarantees of safety and quality are therefore given by the manufacturer of goods that have been independently assessed to the pertaining product standard in accordance with a recognized National, European or International scheme. It is only by recognizing and purchasing such equipment and paying due regard to the required lighting levels that installers can fit a security or safety lighting system which will fulfil its desired function in every respect.

3 Call and access control

Many call methods exist as a means of drawing someone's attention, or with a view to security or care, or to gain access to a given area. Some call methods achieve a personal response but in other methods the response may be of an electronic/mechanical function, e.g. to permit entry to a required area. In the first instance we will look at a specific call system method used where a number of permanent residents need personal attention.

3.1 Nurse call

Due to increased longevity we expect to see a consistent need for residential homes for the elderly. This brings about a constant requirement for various security systems and protection technologies to cater for the older and infirm person.

It must be recognized in the first instance that two different types of accommodation do exist; the rest home and the nursing home. Some 85 per cent of these forms of accommodation are rest homes. In these homes the residents are not ill. Rest homes are controlled by the Department of Social Security (DSS) and the requirements are quite straightforward. Nursing homes however are committed to offering a higher level of nursing care and are therefore controlled by Local Health Authorities.

In all cases both forms of accommodation require a nurse call system together with emergency lighting and fire protection. The needs for emergency lighting and fire protection are detailed in Chapters 2 and 4 respectively. By law the premises may not be licensed until all these requirements are met in full and approved by all interested authorities. The nurse call system in practice has three main features:

(a) to provide a call for assistance;
(b) to confirm that the call was received;
(c) to ensure that the person making the call receives a visit.

Many different types of nurse call systems that can satisfy these requirements are available. However the most simple is probably the hardwired system, which is in itself extremely flexible, easy to use and very reliable.

Calling devices

To make the call, various types of calling devices are used. These usually consist of push buttons or pull cords which can only be reset by a member of staff who must insert a special key into a reset socket. In the case of push button call points the reset socket is built into the unit, together with a confidence light to confirm that the call has been made. Pull cords which are ceiling mounted require an independently wall mounted reset unit.

Call points can also be used to accommodate tail call units which consist of a button small enough to be held by hand. These are connected by a lead of required length to the call point. Signalling is achieved by pressing the button or alternatively by pulling the tail call unit from the call point.

Indication of the call is made on a master panel which is often complemented by a repeater panel or indeed a number of repeater panels for a larger building. For the small home the master panel is often housed in the ground floor office with one repeater panel on the first floor close to the corridor. Extension sounders and over-door lights for outside location of rooms can also be used. Buildings with large passageways tend to make much use of over-door lights. The systems for residential homes converted from old style Victorian houses tends to have additional repeater panels placed on each landing and no over-door lights. Control panels may be connected to indicate every zone in the system or only the zones in a given area, for instance a corridor from where a resident may want to initiate a signal. The master and repeater panel, on receiving the signal, will display an indicator light to show where the call has been made from and a buzzer will provide an audible signal for the nurse. The staff attend the call to reset the system and if they become distracted on the way to the call the sounder will continue to act as a reminder. Calls can also be made if a fire door is opened, if the doorbell rings or if a telephone is operated.

Hardwired systems

In its simplest form the hardwired system essentially follows a circuit configuration as illustrated in Figure 3.1. There are many different control panels and room units or call points on the market, but the wiring of all these derivatives is, in practice, quite similar.

Calling devices use a standard switch process with all signals being monitored by the master panel and formed into indications on the

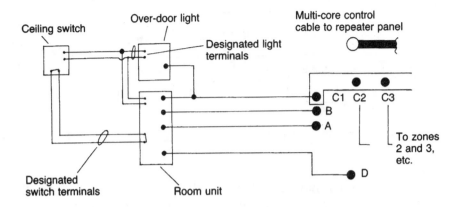

Figure 3.1   *Hardwired system following a circuit configuration*

display panel. Audible and visual signalling devices, such as warning buzzers and over-door lights are then brought into operation.

Wiring is relatively straightforward and since nurse calls operate on low voltage dc with little current consumption, burglar alarm or telephone cable may be employed. If we refer to Figure 3.1, terminals A, B, and D will be identified at both the main control and repeater panel, as well as at the room unit. The zone wire to which the room unit is allocated is shown as C and this is run to the terminal marked C1 at the control or repeater panel. The next zones are run back to C2, C3 and so on. This signal wire is the one which designates where the call is made from and can also be used to indicate a number of call points in a room. These may be connected in parallel on one zone to avoid having to use a master panel with a greater zone capacity. The room unit will feature designated light terminals for over-door lights identified for polarity. It will also have designated switch terminals which are run to devices such as ceiling pull cord switches. In cases where the requirement is to have not more than one room unit on the same zone then a separate C signal wire may be used so that at the panel they are individually identified but they may be wired to a single over-door light.

Extension sounders will have unique terminals and these can be wired to either the main or repeater panel which are themselves connected together by a multi-core control cable. Figure 3.2 shows how the wiring can easily be planned so that by using multi-core cable two zones can operate down five cores and eleven zones can operate down fourteen cores.

When planning any system certain points must be noted prior to running in the control wiring.

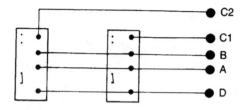

Figure 3.2   *Wiring using a multi-core cable*

1 Every bed must have a call point. However any number of call points in a bedroom may be connected in parallel to one zone.
2 Each room should be on a separate zone.
3 En-suite bathrooms should be included on the same zone.
4 All bathrooms and toilets must have a call point.
5 All seating areas must feature a call point.

These call devices can be wired into a central control unit from where they are fed into a paging system in order to call required personnel. With conventional systems of this type the wearer of a pocket sized radio pager would hear a bleep from the device and he or she should then find a phone to call the operator and receive the appropriate message. If there could be high noise levels in certain areas making it difficult to recognize the bleeping sound of a paging receiver a vibrating unit could be used. This has no audio output but translates the bleep pattern selected at the encoder into a vibration which can be sensed through clothing. Advanced alphanumeric pocket sized radio pagers can convey a visual message following the calling bleep. This appears on an alphanumeric LCD display and minimizes the risks of mis-information or mistakes in situations where the difference could mean life or death. Life saving messages can be conveyed immediately as there is no rush needed to locate a telephone to collect the message. The alphanumeric pager is both direct and time saving. The message, once displayed on the pocket pager, can then be automatically transferred to a memory and could for instance take the form of a telephone number to ring, or a ward to go to immediately in the case of an emergency. A low battery audible or visual warning should be inherent for safety reasons to ensure the correct operation of the pager.

The central control unit monitors all the call devices and each of these can be programmed with the group or individual identity of any pager. Therefore when the call device alarm contact is operated the message is transmitted automatically to the selected receiver or group of receivers. When paging systems of this nature are used in conjunction with an encoder and logging printer, paging traffic can be monitored and analysed to show the most effective allocation of receivers and their grouping. If a printout of calls is generated an assessment of peak periods and particular types of call can be provided and when employed in areas of security, rather than in nursing applications, written details of all calls made can be provided for inspection.

A further type of hardwired system avoiding the use of pagers and that may be employed in the larger installation is the addressable system. This has the advantage of flexibility of installation and use, with the reliability of hard wiring. It uses a network controller or power supply with small unobtrusive LCD displays fitted at any location to provide maximum staff cover. The installation is by a two wire (power and data) protocol network cabling with addressable devices connected to the system along the wiring run. The addressable system suits the needs of specialized homes as the display of calls can be shown wherever required. It also operates in more sophisticated ways, e.g. different urgencies of call, calls shown in different places at different times, sounders operating as quietly as possible etc. Addressable systems will also cater for emergency calls in that staff can make a distinct type of call for extra help by means of emergency tones and pulsing lights and indicators.

The addressable system simplifies wiring in the larger establishment by reducing the number of cables needed and takes into account the greater needs of the user to include such options as 'staff attack'.

The system will operate with power and data on the same wiring pair and additions are made by connecting to this pair. Little planning is required but the network controller is best sited near the centre of the installation to shorten the wiring runs which are called limbs or spines.

A major advantage of the addressable system is that extensions to the system are easily made at a later stage and it has the capacity to have the programming of data from it devices amended as future conditions dictate.

An alternative form of transmission to hardwired or hardwired/ paging systems is the radio system method.

Radio nurse call systems

These systems consist of a base station channel receiver, usually with an LCD alarm readout, printout and clock which is mains operated and complete with a built-in power supply and standby batteries. The

base station receives its messages from appropriate radio transmitters, often in the form of push buttons that may be carried or worn
on a belt. These, when activated, give the alert signal which is
recognized at the base station. Alternatively pendant transmitters can
be worn on a cord around the patient's neck. In other cases the push
button transmitter may be fastened in position in a similar way to a
hardwired component. Pendant transmitters are often used by the
elderly or handicapped living alone or in sheltered housing. Being
worn round the neck, it is activated by pulling it gently against its
cord or lanyard. It may also be worn in the shower.

When the transmitter is activated the base station channel receiver
console will sound to attract attention. It will display the identity code
of the transmitter that has operated and usually produce a hard copy
record of the occurrence, together with the time and date, on its
integral printer.

Since the transmitters communicate with the main control panel via
radio, the installation is quicker and simpler than a hardwired
conventional system. The security of the radio system can be assured
by using coded transmissions, i.e. only transmitters set with exactly
the same 'system code' as that of the receiver will be recognized. By
the use of frequency modulated transmitters and a narrow bandwidth
superhet receiver maximum signal integrity can be achieved. A
greater range is provided by their enhanced sensitivity. In effect any
transmission that is not within a given limit of the centre frequency
will be ignored. Equally any transmission within that very narrow
band but of low power and not possessing exactly the same 'system
code', and lasting for more than a prescribed time duration, will be
classed as invalid. The use of radio systems does in practice extend
well beyond warden type call systems to guard patrol systems, lone
worker schemes and site alarm systems. Great potential does exist for
them in the intruder alarm industry and one can predict major moves
in such a direction. As stated earlier, call systems can be extended to
being used as an electronically controlled means of gaining access to
restricted areas.

3.2 Electronic access control methods

Electronic access control provides a reliable method of ensuring that
only authorized persons can enter a given area and that any exit
routes are not unnecessarily restricted.

Most intruder alarm system control panels have developed to the
extent of replacing keys by keypads. In the electronic access control
field the keypad is widely used to operate the system. In the first

instance it is wise to look at the simple access control system intended for the home or office. This type of system is operated by an externally mounted keypad with an internal control box wired to an electronic door lock. Access is gained by touching one's code into the panel, which is similar in appearance to a push-button telephone keypad. The code, usually some four or five digits, when touched into the keypad will allow access to a given area controlled by this locking mechanism. Alternatively it may open up a speech channel to another telephone or telephone sub-station or base.

A small wall mounted stepdown transformer of 240 V ac input connects the system to the mains via a fused spur and standby power module containing high powered lead acid batteries. The standby module contains power cleaning, regulation and standby charging circuitry to ensure that the battery is always at full charge. In the event of mains failure the integral batteries provide the power source. Such keyless door locking systems should be designed to comply with the relevant requirements of BS 5872 'Specification for Locks and Latches for Doors in Buildings' and feature a set on the latch so that it can remain locked even if the correct code is entered. Systems are available for use with existing cylinder rim locks or mortice latches rather than using the standard lock provided in kit form for use with new doors.

A battery low warning system should be employed and also a digit tamper closedown. This will lock the system if an unauthorized person attempts to gain entry by selecting random numbers. A digit tamper closedown will come into operation after approximately five minutes and is complemented by a duress signalling facility such as an electronic sounder which can draw attention to the area. Once the correct code has been entered into the keypad the electrical lock will energize and become unlocked for a preset time.

In cases of access control where vandalism, corrosion and the weather are likely to cause problems, systems that actually operate through the door may be employed. These systems can be used to operate through any non-magnetic material up to 4 mm thick, i.e. wood, glass, plastic and aluminium. This access control method has all the operational equipment mounted on the secure side of the door or surrounding area including the keypad. A self-adhesive mimic of the keypad is positioned directly in front of the unit on the outside surface. The programmable digit code is magnetically transmitted through the dividing medium to the keypad sensors. Entering the code is achieved by the use of small magnetic keys, hence the security level of conventional keypad access control systems is doubled since the person attempting to gain entry must be aware of the code and must also possess the special small magnetic key. Built-in tamper and duress signalling facilities and variant door entry times are accommodated. Normally 9, 12 or 24 V dc supplies can be used with the command relay suitable for operating most conventional electric locks and strikes. The monitoring and control of restricted areas from a central unit is also achieved by methods of access control.

For applications in industry and commerce, banks, hospitals and other areas where economical access for doors, turnstiles or barriers is needed then access is generally achieved by using an access card. The access card, which contains a specific code, is similar in size to the credit card. This system utilizes a 'reader' to assess and identify the card from stored data and, if valid, the door will be electro-mechanically unlocked. For increased security, keypad and card systems can be used in conjunction with one another using a special sequence. This alleviates problems for security caused by lost cards since any card reading must be followed by entering a personal identification number (PIN), the PIN being known only to the card holder and not shown on the card. In such a system a control box or controller is used in conjunction with the card reader which can recognize up to almost one thousand individually encoded cards.

Entry will be allowed or denied to the card holder depending on whether or not the individual card has been programmed into the computer memory for authorized access.

The actual cards contain a unique site facility code and personal identification number using a high coercivity magnetic stripe material which both retains its magnetism and is impervious to the presence of other magnetic cards. Practice also allows the fitting of a door alarm which is deactivated for legitimate entry. An external alarm output permits activation if the door is forced, and a by-pass switch, if fitted, can allow the system to be operated from a remote point such as a reception area.

Programmable stand-alone card readers are often found as part of a two stage design which allows the reader to be mounted where it is easily accessible for cardholder use and the control unit at a point nearby, where it is programmed. The positioning of the control unit should be secure and convenient. The reader could be mounted, for instance, at a parking lot entrance, data centre entry point or main building reception office. These readers are generally flush, surface, post, or custom mounted. Flush mounted units use a face plate which is usually the size of a single gang light switch cover with the reader slot either vertical or horizontal. These are sunk into the wall so only the flush face plate shows. This can be specified in various materials depending upon the application use.

Surface mount devices extend from the mounting surface and post fit units stand at a height convenient for parking applications. Custom mount describes a non-standard installation form in which the reader is mounted on an L bracket or diverse applications such as elevators.

A feature of many programmable stand-alone readers is its log review function. At any given time an authorized party with a 'master card' can bring up the numbers of the last group of cards used in the reader on an LED display. This is useful in data centre control rooms, stock rooms and vault areas since if losses are discovered the last group of cards used in entering the controlled area can be recalled providing a good starting point for an investigation. These units usually have a non-volatile memory (NVM) as power is not needed to maintain the unit's memory. If the user makes changes such as making void, validating or replacing cards these changes are burned into the memory. The control unit as stated earlier is hardwired remote to the reader so that the solid state card reader is the only component of the system exposed to cardholders. If a one card reader configuration only is to be used then the one reader will of necessity be at an entrance. Entrance elsewhere (e.g. at exit points) is denied by gates or other metal obstructions in parking areas, or by some form of door hardware or an exit button in security zones. Two card reader set-ups operate by having one card reader at the entrance and a further reader at the exit; this is useful when exit and entry operations need be monitored.

In essence then a card access system comprises a reader installed in a non-secure position, and a controller mounted in an inaccessible point from which it is wired to a suitable electrical door strike, possibly also with an alarm shunt output. This would allow a door contact motion detector to be wired to the control unit, giving the card holder a set time to open and then close the door and to be recognized at the selected point. If recognition does not occur an output is closed for several seconds triggering an audible warning.

These systems can be extremely flexible when used in security parking lots. They ensure that only one vehicle can enter a lot on a single permit. If two readers are used the system will insist that if a card is used to enter a car it must also be used to exit, hence it cannot be passed back to a following vehicle.

The need to integrate access with digital access must depend on the application in mind and one should not generalize.

A further technique of access verification is becoming more prominent, namely biometrics or physical recognition. This is highly secure and can be used to validate a card or identify an individual. This method is based on the fact that certain aspects of our physical or behavioural characteristics are unique to each individual.

Physical recognition covers the finger, hand or even eye which are read against enrolled information stored on a template. The finger is measured on a pattern recognition which seeks an image match. For the hand its geometry is assessed to include the height, width and distance between the knuckle joints and finger length. The eye is assessed by the iris to obtain a square image of the capillaries at the back of the eye which are unique in formulation.

Behavioural characteristics are compared against logged voice recognition using a synthesizer system, or handwriting can be employed, involving automated recognition of a signature based on a multiplicity of writing characteristics. Alternatively recognition may be done via keystroke dynamics, whereby the person seeking access needs to prove that a keyboard can be used to a predetermined level.

With biometrics, at the stage of enrolment, when the relevant information is stored, the client is asked to select a level at which acceptance is wanted. Thresholds are set for false reject or false accept. For high security a high threshold is set which will prevent unauthorized entry but means there will be a greater incidence of genuine personnel being denied entry.

It remains to say that the level of security with the different means of access can vary enormously and in practice any techniques can be used in combination.

At this point we can overview the subject of electronic access control (EAC) and draw the reader's attention to the word 'credential' which is the term most used in relation to the documents that verify a person's identity. Essentially these credentials refer to cards, tokens and physical patterns. Whereas cards and tokens are presented for authentication, we say that physical patterns are verified. At the stage when the credential is validated access is granted.

A credential is secure if it is capable of resisting alteration or forgery, but to increase authentication a single access transaction requires a dual process or multi-step verification process combining two or more technologies. This may be through a PIN, biometric feature or even photo/video identification.

Access levels do require different levels of security; an example, external doors need greater validation than internal doors. Therefore once inside the perimeter of a protected premises persons can move from one area to another with a lower level of checking. In practice the most common cards or tokens in use worldwide are based on Wiegand, magnetic stripe and proximity technology in conjunction with PINs.

The most commonly adopted credential types are listed below:

Keypads	Wiegand cards
Bar-coded cards/devices	Infra red cards
Barium ferrite cards	Optical cards

Hollerith
Magnetic stripe cards
Smart cards

Proximity/hands free
Biometrics
Dual process/mixed technology

- Keypads. These are keyboards that are small in size and available in different forms to suit the environment. Durable and inexpensive, they are time honoured but the user must take care when using the system to stop observers from recognizing the PIN.
- Bar-coded cards/devices. Bar coding is often used with other technologies since it is relatively easy to duplicate. The bar code is seen as a set of parallel thin black lines interposed with thicker lines. The lines form a pattern which is interpreted by a reader as a code number. The bar codes are easily printed on cards or other tokens or devices.
- Barium ferrite cards. These cards are very secure and not easy to duplicate. Encoding must be performed at the manufacturing premises and they cannot be erased or distorted or wear out. They are magnetically encoded and are read by a programme cartridge containing a precoded array of magnetized spots. Sometimes they are referred to as BaFe cards.
- Hollerith. For low security applications, these are cards with a technology based on punched holes through which a pattern of light or electrical current can pass. The basic hole pattern can be easily changed.
- Magnetic stripe cards. These are the most popular type. They are not expensive in relation to some technologies and carry a reasonable amount of alphanumeric data that can be encoded at the user's site. Often used with keypads, they have coercive force ratings to indicate the strength of the magnetic force needed to erase the magnetic material: low ratings indicate that the magnetic material is not difficult to erase while high ratings mean that the material is more protected from stray magnetic fields.
- Smart cards. These are available on one of three forms: memory only, memory with hardwired logic, and microcomputer system with a processor. Smart cards have a computer embedded within them to carry user-created information and encryption.
- Wiegand cards. Extremely popular and very secure, these cards involve technology that transforms a metal alloy wire into a bistable magnetic system. The action of this causes a change in polarities based on magnetic attraction to produce a discrete electrical signal pulse.
- Infra red cards. In practice these use a bar code technology, but with the bar code being invisible to the human eye and embedded in the card. They have a long life and are unaffected by magnetic fields. It

is difficult to duplicate them and encoding must be performed at the manufacturer's premises.

- Optical cards. These are extremely secure and they carry huge banks of data in the same way as CD-ROM disks.
- Proximity/hands free. Becoming increasingly popular, these transmit the information by radio waves to the reader. They may be active or passive. Active devices can transmit over greater distances.
- Biometrics. These devices are for high security applications and they associate unique physical attributes with personal identity.
- Dual process/mixed technology. Dual process is the use of a number of techniques simultaneously to verify entry whilst mixed technology is the practice of adopting several card credential methods on one card. Mixed technology makes retrofitting of older systems more cost-effective and also helps to reduce the number of cards that a person may need in one premises or facility. A variant high security combined access method is audio digital.

3.3 Audio digital and intercoms

An extension to digital access control is combined audio digital access control which features a door phone for visitors.

This access control system comprises an access control unit, which can be installed in a secure area, and a combined keyboard and audio unit, which is mounted at the controlled entrance point. A house phone is installed in a reception area or control room, or indeed in the lounge or rest room of a house.

An access code is entered at the keyboard to give normal access. The audio unit is used when a visitor who does not know the access code arrives at the entrance and pushes a call button. The house phone gives an audible indication of the visitor's presence and the receptionist or occupant is then in a position to talk to the visitor. If the visitor is to be allowed to enter the electric door locking system can be operated remotely by pressing a button on the house phone.

Used in conjunction with access control systems, monitor alarms are designed to generate an alarm condition if a door is left in an open position or is jammed open. A monitor alarm is operated by a security switch contact at the point being monitored. When the contact is opened an adjustable period is initiated. If the contact is closed before the time period has expired no alarm condition is generated but the alarm resets for any following contact operation. However, if the time period is exceeded then an audible alarm will follow. This may be silenced by closing the appropriate contact.

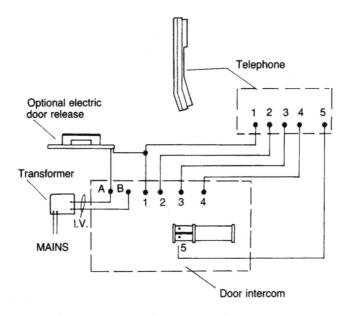

Figure 3.3 *Audio digital and intercoms: typical layout and wiring method*

Access control in practice is a progression from a simple intercom system where calls can be made via telephones or wall mounted door intercoms to a number of stations. Buzzers are used to gain attention by pressing a call button on the telephone and/or the door intercom. These systems have simple exclusive wiring in low voltage signal cable using numbered or lettered terminals. Mains wiring must then be run to the transformer. A typical layout and wiring method is shown in Figure 3.3.

Also available are FM intercoms which are mains operated and transmit and receive through the mains wiring with no interwiring necessary. Each unit is plugged into a 240 V ac mains socket outlet. These do in practice produce high quality sound without any interference from the mains. Units can also be locked in transit mode, so if they are placed, for example, in a baby's room they will transmit to any other unit on the same channel allowing signals to be heard anywhere in the house.

Hands free

The hands free access control system permits entry only to those wearing an authorization pass, but can also be adopted for automatic registration of persons in and out of particular locations. Data such as

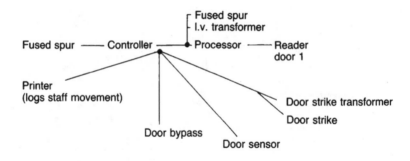

Figure 3.4   *Hands free access control system: configuration for one door*

time, pass number, entrance number and so on can also be recorded. Unlike using conventional readers which rely on the inserting of cards to recognize an individual's presence, hands free uses two forms of passive reader that can be hidden in walls or doors, etc. An aerial can be used, or a proximity reader that has a special housing suitable for environments hostile by the presence of gases, dust or flyings. Hands free can not only be used for regular forms of entrance doors and turnstiles but can be extended for the disabled in hospitals.

Based on a highly secure data transmission method, a badge worn by authorized persons is recognized and gives a command to open the door and record the transaction. A printer can be used to log staff movements. The reader radiates a carrier wave which is generated in the processor and provides all the electrical information, energy and communication with the badge or with a pass fitted on an object such as a key ring. The receiver in the processor acknowledges a pass code. The signal is processed in the receiver and passed on to the microcomputer which is also located in the processor. This then either gives a door release condition, denies access or can be used to generate an alarm. The microcomputer itself can be programmed at times to suit the user. Figure 3.4 shows the system configuration for one door.

We have now looked at how access can be gained to an area controlled by an electric locking system by means of touching in a code or PIN or by entering an authorized identification card into a reader. In the case of hands free methods mention was made of a printer being used to log staff movements. Progressing from this idea, further systems are available that can operate more than one lock. Entry is allowed by one of the aforementioned methods or by a combination of two in sequence.

Printers may then be connected to the system to enable a check to be kept on all entries through the doors concerned. This is done by printouts that state time and date, as well as the code numbers of cards and passes used. Because security requirements may vary according to the time of day and day of the week an integral timer can allow the setting of times for which the authorized access method is necessary. Locks may then be left open or can be completely closed at specific times. By the use of time zones particular programming methods can achieve the correct level of protection of the area for half day working and public holidays. It follows, therefore, that staff movements can, if desired, be logged only at particular times, whilst at other times all persons are allowed free access to specific points.

Such systems clearly identify the location of persons at any given time and give high levels of protection and security. The next stage is to consider systems in which a visual image of the person is also incorporated.

Video intercommunication

An advanced form of access control is the video intercommunication system. This method can be used where an image of the caller is required before one has confidence to permit entry.

The pressing of a call push button on an entrance panel switches on a lamp to illuminate an area scanned by a monitor camera. This signals an indoor buzzer for audible warning and it gives a 'stand by' condition on the monitor, with the caller's image appearing on the screen. There is a large variety of screen sizes available from 50 mm flatscreens upwards. A telephone receiver is then raised to make communication possible. If required the door may then be electrically unlocked.

A push button switch can automatically disconnect the telephone so vision is still available and depression of a further push button can disconnect the monitor. A mute button, if depressed on the house monitor, ensures that the caller cannot hear the conversation in the house. There is a large selection of these video entry security intercoms available and the wiring is very straightforward through a low voltage dc power supply which is connected via the mains. Indeed between the video door station and the room station which houses the video monitor a straight pair of conductors can be used. This means that the original door bell can be disconnected and its wiring used (Figure 3.5).

In cases where there is not sufficient room to mount a complete video entrance station, a TV camera can be mounted independently. The principle of seeing the image of a caller is well known by the more time honoured closed circuit television method.

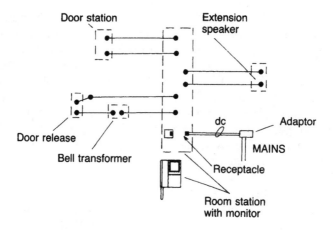

Figure 3.5   *Video intercommunication wiring*

3.4 Closed circuit television systems (CCTV)

Intruder alarm systems are essential protection for all sites that may be at risk, however the greatest possible deterrent to potential offenders is the threat of positive identification. For this reason more sites and businesses such as shops, banks, building societies and similar cash targets with substantial takings are adopting surveillance techniques. This enables photographs of burglars or con-men to be taken during the time of the crime act. Unfortunately without the evidence provided by photographs it is difficult for the police to obtain satisfactory descriptions of offenders from onlookers.

The surveillance cameras used for this purpose fall into two main categories: closed circuit television and photographic 'demand' cameras. CCTV is most useful for continuous surveillance of part or parts of a building or for an outdoor area. Observation is carried out from a central control point using monitors which can survey varying parts of the site or building. For maximum security CCTV systems can be operated continuously but in other cases they can be triggered by an intruder alarm which will only activate them for a given period of time, the scenes can then be stored on video tape if wished.

For specific visual surveillance using cameras and TV monitoring screens, CCTV has become a well-established element. The role of CCTV is wide ranging in practice but it differs greatly in application depending on the area to be scanned.

In areas such as shopping arcades and precincts the public have free access to a large area but, in view of the fact that shoplifting or attacks on the public can occur, some form of surveillance is necessary. Despite the potential benefits to the public, people generally are still not keen on being watched by cameras whilst shopping, therefore in certain locations surveillance equipment must be concealed. Unfortunately this has no deterrent effect, so different premises need different forms of surveillance. The decision whether to use concealed or overt working cameras must therefore be taken by the client.

Industrial sites and offices tend to have restricted means of access and visitors or staff members are therefore unlikely to object to screening by a visible camera. They may even appreciate the reduced risk of theft and vandalism, or indeed attack. The aesthetic value tends to be of less importance in the working environment, so the size of the surveillance system is not crucial. However, in the shopping area situation, where the public do not want to be monitored by large surveillance type equipment, systems must be discrete and hence smaller units are used. Nevertheless, this is not difficult to achieve since in a shopping arcade all surveillance cameras will be sited indoors and many small, light, unobtrusive units are available. Indeed many shoppers would have difficulty distinguishing a surveillance camera from a standard light fitting.

CCTV units used externally must take on a different guise. They need weatherproof enclosures, wiper blades and sunshades, and must be designed with wind loadings taken into account. In shopping complexes they are often found covering the access behind a shop. Shopping precincts have large open public areas, loading bays and car parks. The actual monitoring point should not be on view to the public nor in any area open to the public, as this would allow potential offenders to watch the routine movements of security staff. On the other hand, the monitoring point must not be too far from the shops since the time for security staff to respond would be unnecessarily extended.

The charged coupled device (CCD) camera is a term that is used regularly in the field of closed circuit television cameras. In contrast to the well-recognized CCTV camera, the CCD unit is extremely small and it is not a 'tubed' camera. It uses a silicon chip consisting of an array of MOS sensors coupled to allow the controlled movement of charges through the semiconductor substrate to produce its picture electronically. The CCD camera allows the CCTV to be miniaturized allowing security systems to become very unobtrusive.

CCD cameras are highly reliable. This is a particularly valuable feature for they are frequently mounted in almost inaccessible places. CCD cameras also require much less regular servicing than conventional tube cameras. The CCD camera has low energy consumption, effectively no warm-up time and instant image at switch on. CCD cameras are very much insensitive to magnetic fields which allows their use near heavy current conductors and electromagnetic devices. 'Demand' cameras are more easy to install than constant surveillance CCTV units as they use standard alarm cabling to interface the camera with an alarm system or they can be wired for operation by a simple switch or panic button. In their basic form they are operated by staff when the need arises, i.e. the camera is focused on a specific area and then triggered by a staff member. A 'demand' system is therefore better for sensitive areas where constant monitoring would be undesirable. In other situations demand cameras can be used in combination with CCTV: the demand camera can be triggered by someone watching a CCTV monitor, to provide a high quality image and permanent event picture. Demand cameras can also be operated by a switch used in conjunction with a cash drawer, radio equipment or by means of a timer circuit to take pictures at prescribed time durations.

The extent of the CCTV industry cannot be underestimated but despite its complexities there are certain values that apply across the spectrum. This allows us to make a number of recognized observations.

CCTV is essentially for continuous monitoring or observation only. If it is also necessary to identify and track a target a colour system should be adopted, for observing only and if lighting is poor, monochrome may well be adequate. This is because colour cameras require more light than black and white since the latter type utilize the infra red in the environment. The lenses used in the cameras together with the camera type are related to the security coverage level as summarized below:

Small area coverage: Room, entry hall, doorway.
                          Standard resolution camera and wide angle
                          lens, e.g. 3–4 mm.
Medium area coverage: Small car park, large hall, shop sales floor.
                          High resolution camera to resolve small
                          background detail with medium angle lens,
                          e.g. 6–8 mm.
Large area coverage: Wide spaces.
                          High resolution camera with zoom lens to
                          zoom into detail or view a wider field.
Specific detail cover: To identify faces/cars.
                          Two cameras are needed – a telephoto lens,
                          12 mm plus, and a wide angle lens.
Blind spots: Area under camera.

The more narrow the lens angle the greater the blind spot. Mount the camera as far as practical from target and horizontal.

The camera itself may be low voltage or mains. Generally internal cameras are low voltage and externally they are mains powered. If the distance is great a dc voltage will suffer voltage drops so a 24 Vac system may be superior with the power supply close to the control equipment for ease of service.

The lens iris may be manual, automatic or zoom.

Manual: Mainly suited to internal applications as light variations are low. These are used on cameras with electronic shutters.
Automatic: Superior for external duty coping better with light changes.
Zoom/auto iris: As for automatic but with a zoom facility to enlarge the target in the picture.

The focal length of the lens can be established by entering the statistics of distance to the target, the height of the target and format of the camera into a lens calculator.

Allied to the system will be the multiplexer which organizes the recording sequence. In practice it is a compromise between a video recorder for each camera on a system and a basic video switcher. It records and plays back images and when live displays current images. These are of a simplex or duplex form. Simplex will only perform one key function at a time whilst duplex will perform two, for instance viewing live scenes whilst multiplex recording. The video cassette recorder (VCR) must have good tape management with recorded pictures of evidential quality for prosecution purposes. S-VHS can increase resolution by 60 per cent minimum and is compatible with high resolution colour cameras. For longer recording periods time-lapse capacity machines are used with recordings up to 960 hours.

The remote control of these functions together with the cameras is termed telemetry. New systems are touch screen computer based or menu driven consoles. These offer real text descriptions and site plans. They modernise joystick camera control to provide track-ball devices which are easier to use, less prone to damage and quicker to respond. They link with video motion detection (VMD) which is intended to capture images only when an unusual event occurs.

Of course the extent to which CCTV is employed is also governed by financial allowances and must be balanced against the duty it is intended to fulfil.

Passive infra red security camera systems

As a low cost alternative to CCTV systems, PIR security cameras utilizing the latest technology can be considered. PIR cameras have high immunity to false alarms and are suitable for use in domestic, commercial and industrial applications. Such devices use an instant photograph principle and are housed in a reinforced enclosure with a PIR detector and associated circuitry.

They may be used as a direct alternative to a standard PIR detector so that in the event of an intruder being sensed the security camera triggers the remote audible alarm and simultaneously photographs the intruder.

A variant application is to use the PIR camera unit as a stand-alone device with a power pack. In this case the camera is armed remotely and will take a photograph of any intruder entering a detection zone. In other applications the camera may be trigged by an external signal from a recognized sensing device such as a PIR, ultrasonic microwave or capacitance detector. Pressure sensitive mats or reed switches can also fulfil this function.

Multi-shot capability is normally available and the replacement of expensive film cartridges used in existing time based security cameras is eliminated. Also the manual monitoring associated with CCTV systems is not required.

In the main, PIR security cameras are in robust weatherproof housings which are designed to withstand any attack; they also should have LEDs for walk test and film exposure practices.

The monitored zone of PIR security cameras differs from model to model. This must be studied before installation. However, it is generally found that the total detection range will be in the order of 8–10 m.

Slow scan/telephone video transmission

It must be clear that one can never be certain if an alarm is false unless the scene in question can be viewed, and this philosophy applies even when the most sophisticated alarm sensors and equipment are used. Hence if pictures of the scene can be seen at the same time as an alarm occurs then correct information is gained. Slow scan television (SSTV) is effective in this way by using fully automatic camera sequencing. Slow scan is a form of video phone as it sends pictures over very narrow frequency bands in the audio frequency or speech range, this is performed over normal telephone lines by segmenting the pictures (wide video frequency spectrum) down into small portions and then reconstituting them in the receiver equipment. This results in a series of still pictures not unlike a slide projector image. The term 'slow scan' is derived from the time taken to transmit pictures in a segmented form over the telephone network in a serial manner. It is often also known as telephone video transmission.

If automatic dialling and answering modules are used the telephone is only put to use whilst the system operates, hence manning costs only cover the telephone line call as and when used.

In security applications slow scan can be seen as a low cost method of monitoring and providing surveillance over very long distances. Since the cause of alarm is verified by an appropriate monitor in a receiving station no false alarm is ever conveyed to the police so there is great credibility. Although numerous SSTV systems exist and have common principles it will be found that differences do occur in features, operation and presentation.

Audio communication facilities, if provided, enable the operator at the receiving station to be in speech contact with the site. He can even use the public address system as a warning deterrent to potential intruders who may have been detected. If telephone line failure occurs the alarm pictures are held in a memory until the line is restored and if the mains power should fail or be disconnected a back-up battery source will maintain the picture store.

In view of the fact that the telephone network is used these systems are also capable of operating via fibre optic trunks, microwave links and cellular radio systems. In addition slow scan can be operated in conjunction with CCTV and intruder detector methods. It is of great value when protecting remote country depots where animals can cause movement detectors to activate, and equally it is useful when protecting sites in close proximity to residential areas, where audible intruder alarms cause complaints from neighbours, especially if they are a regular occurrence.

In real terms the actual installation of CCTV is a straightforward process, however a professional system design is essential if the required monitoring is to be achieved. Nevertheless for the less complex and small installation many surveillance system kits are available. These are becoming very popular and are worth special mention. They consist of a CCD camera with cable and connectors and a monitor with a built-in switcher (usually 4 way). The camera itself is line fed, in that it receives its power along the same cable that is used to send its picture back to the monitor. The switcher is a device built into the monitor, enabling the user to select which camera is to be projected onto the screen, since it is possible to connect additional cameras.

As a progression from this one can elect to install a system by purchasing a monitor, separate switcher and a specific number of

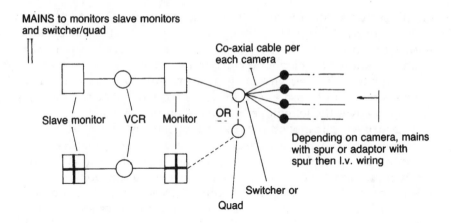

Figure 3.6 *Surveillance system*

cameras. These cameras may be mains powered or work from a low voltage dc supply via adaptors. An option to using a switcher is to adopt a quad splitter or multiplexer. These use digital techniques to compress four full pictures onto one monitor screen for the former or an even greater number of pictures for the latter. The wiring is shown in Figure 3.6 and involves mains connection by standard three pin plugs to mains socket outlets for the monitors, video cassette recorder (VCR) and switcher or quad/multiplexer. The cameras are connected to the mains by fused spurs or to low voltage dc adaptors via fused spurs to the mains. The cabling is otherwise co-axial with a separate cable for each camera to the quad/multiplexer but a single cable from there to the monitors and video cassette recorder.

A slave monitor enables an interested party to view proceedings from a remote point, whereas the operator of the switcher or quad splitter/multiplexer (for multi-screen surveillance) can be closer to hand and watch his monitor from a local point. It is also possible to incorporate a time lapse VCR with a date and time generator to record events for later viewing. The latter VCR type offers up to 24 hours of video and audio recording on a single tape by recording frames with pauses between them, logging the date and time of all activities. These systems are sometimes referred to as CCVE – derived from 'closed circuit video equipment'. Certainly the benefits of all surveillance systems are great, from the small installation through to the major network. These visual methods also allow verification of alarms, so the market can surely be expected to expand in this direction in the future.

3.5 Fence detection systems

Fence detection system methods, which give first line perimeter protection, are generally used with support cameras. With all the systems discussed so far we have looked at how maximum electronic protection and security can be achieved with minimum false alarms. However, with fence detection techniques this aim of no false alarm signals does not appear to be possible, so verification with cameras is usually incorporated. The environmental problems involving heavy passing traffic, large dogs or innocent persons in the vicinity of the fence will always cause a problem so in real terms these detection systems are only applied to complement an existing perimeter fence.

The first systems to consider are those that have sensor cables installed on the fence fabric to detect both attempts to scale or to cut through the fence. The number of disturbances within a time interval that the system can tolerate before generating an alarm can be selected and changed by the system user. Different channels can also be adjusted separately so that sensitivity levels, to prevent scaling of the fence or cutting through sections of it may be selected as appropriate. The sensor cable in fact acts as a tuned microphone listening for disturbance of the fabric but it is apparent that the fence sensor must be fitted on the side where people cannot legitimately be, otherwise a fence sensor system within a physical fence is needed.

Microwave beam fences are usually found close to and within perimeter boundary fences. These provide a secure area, up to a distance of some 200 m, using a transmitter and receiver principle. They should be installed on level ground and are best employed on long straight boundaries as essentially they use beam breaking techniques. In cases where there are ground undulations and one is concerned that an intruder could evade the beams by crawling under them, then beam shaping or phase sensing can be applied using vertical or horizontal aerials to give increased ground cover or extended ground and high-level cover. The specific way of installing these aerials would be in the data provided by the manufacturer for the equipment.

An alternative to the microwave system is doppler radar. This is based on microwave principles but has two beam patterns, wide angled and long ranged. The beams are applied where movement would normally be towards them, such as down a path, since doppler radar uses a single head which both emits and analyses the signal in doppler shift.

We have already looked at how infra red beams or passive infra red detection can be adopted to activate lighting and both are used at times in perimeter fence systems, even though their range is so much smaller than microwave. The other methods warranting a mention are those that are buried underground to stop digging, or electrified security fences which comprise of a steel wire grid. Across this grid high voltage short duration pulses are applied and then monitored for integrity by a control unit.

In practical terms all fence detection methods are not intended to give immediate high output audible alarms but are designed to be linked into closed circuit television camera systems. Using a digital technology format, when the alarm is generated, the process presents a picture of the area where the signal originated and this is incorporated into a video recording. Adjustments to the sensitivity of any particular detection device over a period of time can always help to minimize sporadic operation. Visual confirmation of the activation of the detector is essential.

3.6 Barriers

It remains the case that fence detection systems give protection to an area as a perimeter defence but tend to be used with cameras to verify an alarm signal. Barriers are, of course, used with all access control systems and although we have briefly mentioned a few types throughout our observations on access control and video intercommunications, we can conclude the chapter with an appraisal of the most used barrier versions with locking and unlocking facilities.

There are two conditions of electrical locking:

● Fail secure. In this condition the lock remains locked when power is removed from the device.
● Fail safe. The lock becomes unlocked in the event of the power being off.

The four common lock types in use are electromagnetic, electric strikes, electric locksets and electric deadbolts. We will consider each in turn.

Electromagnetic locks

These are fail safe and operate by magnetic force as they consist of an electromagnet and a strike plate. The electromagnet is installed on the door frame and when energized it attracts the strike plate. They may be direct hold or concealed. The direct hold version is mounted on the secure side of the door and frame. It is installed on the surface and so is not concealed. The shear type is embedded and so is concealed within the frame and the door.

There are different grades for electromagnetic locks depending upon their holding force.

With all electromagnets a sensor is needed to report the status of the door as being open or closed. The current consumption of the electromagnet depends upon the grade and they often come complete with a power supply specific to the unit.

Electric strikes

These are the most popular locking device and are a standard product with small intercom systems. They may be fail safe or fail secure and are available in a variety of voltage ratings. In practice they are able to replace a large proportion of existing mechanical locks and so are easily added to an original application.

The strike is mounted in the frame, either on the surface or concealed, and is available with accessories for inward or outward opening doors. In operation the strike has an integral pocket that collapses or rotates, enabling the bolt or latch of the door to be freed or escape from the keeper pocket.

The electric strike can accommodate either latch bolts or deadbolts. The latch bolt is a spring-loaded bevelled latch that slides over the keeper pocket when the door shuts and then springs back to a locking position. The deadbolt is not bevelled so the latch is always either protruding or retracted.

Customers often prefer ac devices as the sound of the unit operating is apparent when they are attempting to open the door and realize the lock is energizing to permit entry.

Electric locksets

These are similar to a mechanical lockset but the mechanical operation of a standard key is replaced by an electrically operated system. Once again they may be fail safe or fail secure.

The most popular versions are the cylindrical lockset and the mortise lockset. The former has a door handle on each side of the door and the latter a time-honoured mortise lock embedded at the door edge.

With these locks the electrical power must be accommodated at the location of the lock on the door and this presents certain complications. The cabling is achieved by a flexible door loop which must be provided on the secure side of the door. This carries the supply from the fixed wiring at the frame to the door itself. The cabling must then run to the lock and is best concealed within the door itself. The electric lockset is often found on double doors.

Electric deadbolts

These are often found on double doors and on doors which swing in two directions. The comprise an electrically driven bolt which protrudes when energized. The deadbolt may be found fitted either on the door or the frame. It does not have a collapsing or spring action. They are found both as fail safe or fail secure variants and may be used as sets with multiple units protecting a door or opening.

One requirement which applies for all doors controlled by an electronic access system, is that they come complete with a door closing mechanism. In addition they should have a sensor or switch to monitor the door for being correctly closed. We should add that a means to exit the door must also be provided by an opening handle or push to exit button (egress) to give an opening electrical pulse.

Gates

The next barrier form to consider is the gate. Almost any gate can be electrically automated but in practice this is only employed to delay unauthorized access to vehicles, thus allowing pedestrians to walk around them or though an adjacent gate. These automated gates often have two-way speaker systems and must come complete with sensors to ensure that vehicles cannot be impacted by the movement of the gate and that the exit is clear for the vehicle once the barrier opens.

The typical gate forms involve a pivot action or pop up, or they can be pulled back on rollers in channels.

For full vehicle control any credentials can be used to give access. In those cases when a need for more rapid traffic flow exists, authorized vehicles can be fitted with a proximity token or bar code reader, as these are effective at distances of the order of 10m. To be highly secure these barriers need constant surveillance.

Internal gates to control pedestrian access are more specialist. They may be of a turnstile or lane/alley form. In high security applications such as financial institutions or government buildings, mantrap vestibules are used. These have holding areas in which those attempting to gain access are held within two doors that form a secure point of monitoring.

It is appropriate to complete our overview of barriers by stressing that we must never lose sight of the fact that building and fire regulations also bring needs that must be satisfied with regard to emergency exit routes. The requirements differ depending on the building type and use but essentially these exits and doors must:

● be clearly visible;
● be simple to open and operate;
● be designed with the minimum amount of operating hardware;
● not be locked internally;
● fail safe electrically;
● have door closers to effect closing when released;
● be manufactured from fire-resistant materials.

The access control system must therefore also be governed by these needs in addition to general exit release procedures.

3.7 Integrated systems

The reader will have come to appreciate that very often the different types of security systems can be made to complement each other. In practice any two systems can be integrated. This is the art of combining diverse elements into one whole or collective central control.

The joining of different systems can allow the sharing of information. Integration can be seen as a system that covers all possible security as well as building control applications. It may be as simple as connecting an alarm panel to an access control system to stop personnel entering an area when that area is armed, as depicted in Figure 3.7. Alternatively it can be as complex as having a single point of administration for all the systems in a building, as depicted in Figure 3.8.

The reasons for integration involve the advantages of a single administration point with one network, allowing tailored systems. It is clear that existing systems may look similar if they have Microsoft

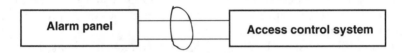

A pair of wires from the alarm panel output to the door controller that secures access for that area

Figure 3.7 *Intruder/access integration*

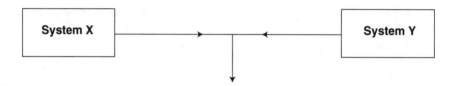

Software created so that the user interface reflects all systems that exist and how they have been installed

Figure 3.8 *Single point administration*

Windows based software as they have the same language and understanding of the facilities, but they need to talk to each other and this is best designed in at the point of initial product specification. For instance, although Windows 95 and NT software offer similar graphical user interfaces, extra intelligence must be added between the system to translate the messages from one system into actions in the other.

A major consideration is that if a failure occurs with an integrated system it must be possible to be able to determine which of the separate systems has the fault. If special software is linking two systems it is advisable to be able to manage the systems separately with individual maintenance. The options for integrated systems are, therefore, to have only discrete systems with simple interconnections via standard electrical input/outputs or to use sophisticated systems with bespoke software. Certainly the control of one or more large site complexes from a central point is an attractive proposition but it can only be achieved by using products geared to integration with intelligent gateways from product to product.

At this point in time many intruder alarm control panels do have a circuit designated for a smoke detector but this is only of value where a fire alarm is not required by law. To integrate more complex intruder and fire systems does present certain considerations. The security industry uses Inspectorates such as NACOSS which is essential for installation in the approved market whilst the fire industry has LPS1014 which is not compulsory.

Difficulties exist in locating approved smoke and heat detectors compliant with the Loss Prevention Certification Board (LPCB) for 12 V systems. Any fire detection system should be installed to BS 5839 Part 1 for any site that requires a fire detection system by law. This Code of Practice states that all fire control and indicating equipment (CIE) must comply with BS 5839 Part 4, but standard intruder panels will not meet this. An exception where this is not invoked is BS 5839 Part 6 which is the Code of Practice for the design and installation of fire detection in dwellings. In these cases the intruder system can be used as a combined fire and intruder panel. The trend can only be to adopt a combined control panel to comply with LPS 1200 and BS 5839 Part 4 to give a true integrated intruder and fire system. For the future consideration must also be given to EN 54 Part 2 as this is the European Standard that is to take over the role of BS 5839 Part 4.

We can appreciate that with digital processing and communications protocols becoming common knowledge, manufacturers now tend to offer products that make integration easier. Simple integration can be by NO or NC switching, allowing one system to control the other using pairs of wires so that one can signal the other when user-specified conditions occur. This may be via the volt-free contacts of a relay or by a designated ouput. As an example, a CCTV system can be programmed to invoke a particular camera and display the video as an access system alarm. For facilities needing greater control of the CCTV system it is normal to use a CCTV controller with access control integration capabilities.

In summary, we can reiterate that simple integration by a pair of wires is possible with all systems but true integration with all controls being operable from a centralized point with the systems still able to be separately managed must be designed at the initial stages.

3.8 Fibre optic transmission

Having studied integrated systems we can say that fibre optic or optical links are themselves multifunctional. By such means a single cable running around a building can transmit not only alarm signals but also guard intercom signals, CCTV signals and audio signals, as well as logging data from patrol points and lighting system relays and accessing control information.

Fibre optics comprise cores of filaments of extruded glass in a continuous length. This glass is essentially pure with a high transmissivity to light along its length, and is available in many different fibre sizes. The different fibre diameters have differing maximum optical attenuations (dB) with different maximum input power limits (microwatts). The quality of the fibre is defined by a numerical aperture number (NA) which is a measure of its ability to absorb light for transmission purposes. The glass fibre is surrounded by a sheath which provides mechanical protection for the fibre and is usually referred to as a jacket.

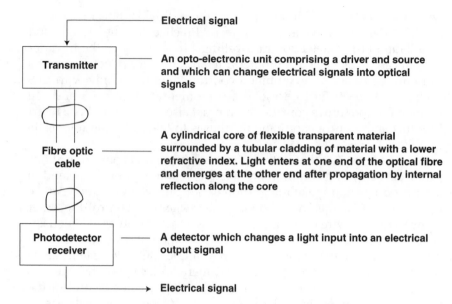

Figure 3.9   *Fibre optic transmission*

In practice a fibre optic transmission system will be as depicted at Figure 3.9, comprising a transmitter, fibre optic cable, a photodetector/receiver and perhaps a reception amplifier.

In operation the transmitter accepts an input voltage signal from a system and then transforms this signal into a modulated beam of light. This light is then focused within the fibre optic cable, being sent at the speed of light to the photodetector/receiver which in turn converts the light signal back to an electrical signal by means of an amplifier. The frequency of transmission and bandwidth vary, but as examples for CCTV are in the range 10 Hz to 10 MHz and for audio systems are in the range 10–20 Hz. Bandwidth is actually the difference between the upper and lower frequencies which can be sent along a communication channel. When used in radio engineering bandwidth is quoted in cycles per second (Hz) but the word bandwidth is now also used to specify digital data transmission channels, so that Hz then refers to bits per second.

These links are particularly useful for transmissions between buildings to avoid risk of lightning damage and EMI or noise that could be generated by stray electromagnetic fields. Also, when using fibre optic cables there is no need to be concerned with electrical isolation.

There is a growing recognition of fibre optics within the security industry and they are now used extensively in CCTV applications, particularly in harsh environments and in long runs where the cable can be buried as a multiplexed medium to a monitoring point.

Advantages of fibre optic links are that all signal transmissions are possible, there is little loss through the cable itself (attenuation) so that up to 5 km of cabling can be used without amplification, and electrical or radiated energy interference is not possible. Equally, there can be no corrosion and the cable can withstand harsh environments and send huge amounts of data and simultaneous transmissions over small cable diameters.

The cost of fibre optic links is a major consideration as it is a more expensive medium than traditional methods. There are limits on bending and care must be exercised in terminating and jointing. In the past the fitting of connectors has involved the use of epoxy resins and ovens to cure the resins, followed by extensive polishing to gain a perfect finish. With new generation equipment, excellent terminations can be made with crimp-type connectors, similar to those found on coaxial cables. This standard crimping process is followed by light polishing.

Fibre optic systems, unlike copper systems, attenuate the signals equally, irrespective of their frequencies, so rather than several controls being used to balance the signal after transmission, simple amplification is used. This is carried out by an automatic process called automatic gain control (AGC). It is performed in the receiver units and the same electronic circuit ensures that the signal level is adjusted throughout the life of the equipment.

The increasing use of optical fibres will inevitably change the method of long haul linking of security systems and the ways of negating the electrical interference that can presently be induced in such systems.

4 Fire alarms

Fire alarm systems can unfortunately not prevent fires but are intended primarily to protect loss of life and to give early warning in the case of a fire. They provide an opportunity to evacuate a burning building and minimize property loss by early notification to the authorities.

One of the worrying areas in respect of fire safety is the increase in arson. It is a sad fact that since arson attracts publicity and attention it does encourage the arsonist to commit further crime. Therefore it is essential to raise general awareness of the painful social consequences of fires and to develop a greater sense of responsibility among property owners and occupiers. Also efforts must be stepped up to teach fire safety in schools to improve fire awareness at an early age. In most commercial, multiple occupancy and industrial buildings, a case can be made for installing fire alarms and it is to these systems that this chapter is devoted.

Fire alarm detection systems fall into two main categories, the basic break-glass or manual fire alarm system, and the automatic detection type.

The manual alarm system, which consists of break-glass call points and alarm sounders wired to a control panel, can only be operated deliberately and the alarm is generated when the break-glass point is activated by a person who has detected a fire incident.

An automatic system, which consists of smoke and heat detectors in conjunction with break-glass units and alarm sounders connected to a control panel, is designed to produce an alarm condition whether or not personnel are on the premises. This gives an early warning of a fire incident.

The design of a system is subject to a large number of variables since much depends on the type of premises to be protected and the specific level of protection needed. Full design criteria for fire alarm and detection systems is given in BS 5839 'Fire Detection and Alarm Systems in Buildings' which details spacing requirements for the various types of detector. However, it is always necessary to consult the local Fire Prevention Officer who can advise on the type of system needed and any local or government legislation that will have to be complied with.

The majority of industrial and commercial establishments do need a fire alarm system to protect life or goods and machinery. Zones must be established throughout the protected premises with consideration being given to accessibility, size and necessary fire routing. In occupied areas every zone must be accessible from main circulation routes leading from the point where the control panel is sited.

Guidelines do exist for zone sizes. In general:

(a) A zone should not extend beyond a single fire compartment as defined by the main walls, floor and ceiling. However if the total floor area is not greater than 300 m² then the building is classed as a single zone, no matter how many floors it has.
(b) The floor area in total for a zone should not exceed 2000 m².
(c) Where stairwells or similar structures extend beyond one floor but are in one fire compartment then the stairwell should be a separate zone.
(d) Any distance travelled within a zone to determine the position of a fire that has been indicated in that zone should not exceed 30 m. To meet this requirement in a zone subdivided into rooms it is useful to fit indicators outside the doors so that it will be apparent that a detector within a room has been operated. These are also helpful if doors are liable to be locked.

4.1 Detection devices

As previously stated fire detectors are available in two forms: manual and automatic.

Manual detectors

The break-glass call point is a manually activated device intended to enable personnel to raise the alarm in the event of a fire by breaking a glass cover on the unit which activates the alarm system. Such devices are available in many forms. The glass cover on some units may be broken by a hammer or other robust object. Alternative call points feature a frangible element in which the operating switch is held off by the edge of the glass. These use the scored glass principle, so that firm thumb pressure is sufficient to break the glass. This type eliminates the risk of false alarms that could be caused by glass elements breaking on their own. Glass fragments are also prevented from entering the switch, stopping it from moving to its operating mode. A protective plastic coating on the glass prevents operator injury and any release of glass fragments. Once commissioned, call points of this type may be tested by means of a special key to ensure correct operation. All call points should be manufactured in a red coloured material and comply with BS 5839 Part 2 'Specification for Manual Call Points' to cater for interior use or as a weatherproof version for outside use where adverse conditions apply.

In hazardous areas special types of break-glass call points are required. Hazardous areas are classified as those that may contain flammable gases, vapours, liquids or solid substances which can either burn slowly or can explode suddenly depending on the substance and prevailing conditions. The petrochemical industry of necessity falls into this category. The operating switch in call points mounted in such areas are of a carefully defined construction since they must themselves be incapable of causing any spark between the switching contacts. Switches for this duty must also meet onerous conditions of construction and electrical insulation properties.

Certification authorities are active in approving items for hazardous areas depending upon the sector of industry for which the call points are intended.

Automatic detectors

Automatic fire detectors must be able to differentiate between a fire and the normal environment existing within the building, therefore careful consideration must be given to the particular area in which the detector is to be sited. The incorrect use of a smoke or heat detector will, without doubt, lead to unwanted alarms. For instance smoke detectors should never be used where there is a likelihood of smoke or fumes being present during the normal working environment, such as in kitchens or boiler rooms. Care must also be exercised in the siting of smoke detectors where there is a possibility of smoke or fumes reaching the detector under normal circumstances. This condition can occur immediately outside a kitchen or bathroom door, near an extractor from a fume laden environment or near an open fire. The main types of smoke detector are the ionization smoke detector and the photoelectric/optical smoke detector.

Ionization smoke detector

An ionization smoke detector responds when smoke, having entered the detector chamber, causes a change in ionization currents within the detector. It will detect smoke particles given off by a fire including invisible particles and is also sensitive to fumes from various chemicals including petrochemicals. This type of detector is best used in buildings in which the fire risk is likely to be an outbreak of flaming, as distinct from smouldering.

Precaution against any nuisance alarm is effected since the detector will not signal an alarm unless the chamber has contained a smoke concentration at or above its threshold for a given period of time – usually in the order of 5 seconds.

Photoelectric/optical smoke detector

A photoelectric/optical smoke detector uses a form of detection where a photoelectric cell responds when light is either absorbed or dispersed by smoke particles. It is the most suitable detector for fires where smouldering is likely to occur, such as bedding or linen and it is also less susceptible to strong air currents.

This device is particularly suited for difficult applications such as elevator shafts where air currents are present and smouldering is likely.

Rate of rise heat detector

In cases where smoke detectors are not acceptable then rate of rise or fixed temperature heat detectors should be employed. The rate of rise heat detector responds to a certain rate of rise in temperature and triggers before a limit temperature is reached, unless the rate of rise is very slow in which case the top temperature element will operate at approximately 60°C.

Fixed temperature heat detector

A fixed temperature heat detector responds only to a fixed temperature usually 60°C or 90°C and is not as sensitive as the rate of rise detectors. They are used were rapid changes in room temperature can occur such as in boiler rooms or kitchens.

When siting detectors it should be understood that the greatest concentration of smoke and heat will occur at the highest parts of enclosed areas, therefore these are the points where the detectors should normally be sited. For heat detectors the element should not be less than 25 mm or more than 150 mm below the ceiling level. Smoke detectors should not be mounted less than 25 mm below the ceiling level or more than 600 mm. In the case of premises with an apex construction, smoke detectors should be installed in each apex.

For simplicity of installation, economy of servicing and quick interchangeability most manufacturers' smoke and heat detectors are installed in a common base. The precision made base has spring contacts which ensure excellent electro-mechanical connection to whichever lead is positioned into the base. The sensor head is usually positioned in the common base by inserting it into keypad slots and then rotating it until it rests in the latched condition. This lock prevents any malicious removal of the detector head by unauthorized parties in low ceiling installations since a special tool is needed to effect removal. However, for installations where the detector is almost inaccessible the mechanical latch can be disabled.

In applications where wall mounting is required for detection devices, for example large span open type interiors, then solid state light source smoke and heat beam detectors may be used. These can be set to respond to both or either smoke and heat by using a photo-electric beam receiver/transmitter method. These units are designed to be installed so that the length of the beam projects some 0.5 m below and parallel to the roof level. When smoke becomes present in the beam the received signal is reduced and if this is within certain limits an alarm is generated. If thermal turbulence above a fire causes modulation for more than a given period of time then the heat alarm output is activated. If these units are to be used in low temperature environments then heaters can be used in conjunction with the device. Both manually and automatically operated detection devices are wired back to a central control panel which should be mounted close to the main entrance of the protected building. The LED indicators on the control panel will enable the emergency fire services to quickly establish the source of the alarm when they enter the building.

The common mounting base of the detection device is also equipped with an LED alarm indicator. Further terminals are often provided for the connection of a remote indicator, which would be used in situations where the indicator on the detector base is not normally visible, i.e. locked rooms, cupboards or ceiling and floor voids.

The fire detection system itself operates from the 240 V ac mains supply using a 24 V dc system voltage with a charger rectifier circuit incorporated in the control panel. In all instances the power supply equipment for a fire alarm system should be exclusive. If the system must be used in computer controlled buildings and be combined with other control or emergency systems the power supply reliability must never be impaired. The mains connection should be made using a fused spur for the sole purpose of the fire alarm system and should be painted red and bear the label 'Fire Alarm. Do Not Switch Off'. The installation should be in full accordance with the current edition of the IEE wiring Regulations but advice must always also be sought from the local Fire Prevention Officer as to the acceptable wiring type. Also the system, when operating from the standby 24 V dc power source, should be capable of working under normal operation for a period 24 hours greater than the maximum period for which the premises are liable to be unattended. The standby batteries must have an expected life of at least 4 years and be capable of operating the sounders for 30 minutes after a certain minimum duration. This minimum duration varies with the type of system and building occupancy.

4.2 Control panels

Residential computerized control panels intended to give total security to a home can provide protection against intruders and fire. These offer peace of mind whether one is at home or away and have separate indication and outputs for fire zones. The fire detectors are wired into 24-hour monitored loops since these are active permanently 24 hours a day, 7 days a week. However, in the commercial or industrial sectors independent intruder alarm and fire alarm control panels are normally used. It will be appreciated that this is necessary because of the more complex nature of the systems employed. For the medium size application many competitively priced fire control panels are available to cater for one or two zones of automatic detection and manual alarms. These should be manufactured to meet the functional requirements of BS 3116 Part 4 'Specification for Automatic Fire Alarm Systems in Buildings Control and Indication Equipment' and BS 5839. They should have an integral charger and be able to accept standby battery power supply in the event of mains failure.

Certain single circuit fire control panels can be extended to two circuits by plugging in a further circuit zone card. These may also be extendable upwards by adding purpose designed zone boxes.

External LED indicators show the presence of mains, correct function of the charger circuitry and also that the battery is not disconnected. Fault indicators are also needed for any zone open, alarm sounder open or short circuited, and alarm silenced.

Fire alarm repeater panels, to match many programmable control panels, can be used to enable fire indicators on the main control panel to be mimicked at another point. This could, for instance, be at an extra building entry point or at a gate house. Essentially the control and indicator panel will be governed by the size of the protected building and the extent of the automatic protection it must provide. This is determined by the equipment specification such as the number of zone circuits, sounders, battery standby, plus any extra control relays or individual customer requirements.

The control panel should be installed in an area of low fire risk, usually on the ground floor by the entrance that the fire services would enter by. This should be an area common to all building users.

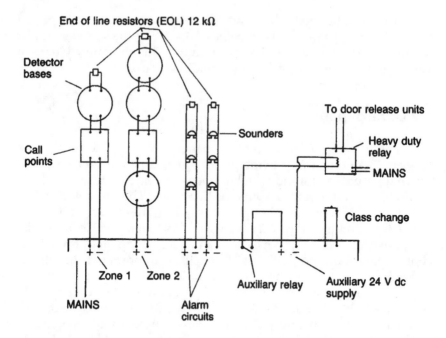

Figure 4.1   *Typical wiring*

This point should also be protected by an automatic detector and have an alarm sounder sited close by. In addition, next to the control unit there should be a zone designation chart or a plan in diagram form detailing the zone locations.

The wiring of a typical system is shown in Figure 4.1. This system is in the open circuit fault monitored mode with 12 kΩ monitoring resistors fitted at the ends of circuits. It will be seen that automatic fire detectors and manual break-glass units and sounders are wired in parallel, are continuous and are without the use of spurs. The class change terminals shown are those which, if shorted, will cause the sounders to operate.

It is also possible to wire a magnetic door release into the system. These can be used with fire systems to allow fire check doors to be kept open under normal circumstances. However, should an alarm condition be generated then the doors will be closed by the necessary self-closing mechanism. These units consist of two essential parts, a metal plate which is secured to the fire check door and a solenoid which is located on the adjacent wall position. A manual release button is also fitted in the side of the release unit. As these must work on mains voltage, a 24 V dc coil rating relay is used as an interface between the fire control panel and the magnetic door release unit. Relays of this nature can also be used to switch other heavy loads on fire alarm systems for smoke ventilators or to effect a plant shutdown.

Figure 4.1 shows typical wiring. Care must be taken to run parallel circuits with regard to polarity and to correctly use the end of line 12 kΩ (EOL) resistors that are provided with the panel. These must be fitted to the last detector on that particular zone and to the last sounder on the alarm loop. The detector bases are wired positive in and then positive out on the corresponding terminals and the same is done for the negative. The call points, which are non-powered switched devices, are shown having a different value resistor within the unit and in series with the normally open contacts. The control panel can recognize short and open circuits in the wiring or detector removal by monitoring these resistances. In some systems continuity diodes are used in automatic detector bases. These are connected so that they are shorted out by the head when it is in place. A path for the current in the loop when the head is removed is therefore provided by the diode.

The control panel will also recognize short and open circuits in the alarm sounder loop and will give the necessary output depending on the state in which it finds the components (Figure 4.2).

4.3 Signalling equipment

In any building the number of fire alarm sounders must be at least two. They may be either a bell or an electronic sounder. In all cases the alarm sounders must have a similar noise tone and be different from sounders used for other purposes, such as intruder alarms.

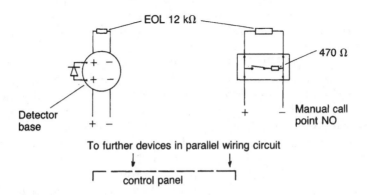

Figure 4.2 *Detector and call point wiring method*

Sounders should be wired on a minimum of a least two separate circuits, so that failure of any one circuit cannot cause a shutdown of all warning devices in a building. The level of sound must never be so high as to cause permanent hearing damage and a larger number of quieter sounders are advocated rather than a few high volume sounders in order to prevent noise levels in certain areas becoming too loud. However one sounder per fire compartment is always necessary since it is unlikely that noise levels will be satisfactory if there are dividing walls. For premises where sleeping persons must be awakened then the sound level should be 75 dBA minimum at the bedhead – this applies in buildings such as hotels and boarding houses. Otherwise a minimum sound level of either 65 dBA or 5 dBA above background noise should be produced.

Many types of weatherproof electronic sounder are available and they may be used with confidence in outdoor locations. A volume control inside the unit, when fitted, enables the sound output to be set to suit varying locations. Flame retardant housing versions are also available for use in industrial applications.

When a visual alarm is also required because the area itself could have a high ambient noise level, high efficiency 24 V strobe lights can be used in conjunction with the audible alarm.

In practice the satisfactory wiring of the fire alarm depends very much on the interconnection of all the components. It will also be appreciated that certain connections may have to function as intended after a significant period of time and even after being subject to fire. The cables concerned are those pertaining to power supplies, control equipment and the sounders. The cables to automatic detectors and call points, etc. must of necessity function to trigger an alarm but are not needed once the alarm has sounded. Cables are therefore classed in two groups. Group 1 is for cables to manual call points, smoke and heat detectors, whilst Group 2 covers those cables essential to the operation during the fire, such as those to the sounders, power supplies and the control equipment.

The British Standard, although it emphasizes mineral insulated cables, also accepts many other types, but there may be restrictions on the use of some, unless they are given special protection. Methods of protection include burying them in walls with a plaster cover applied, or shielding them with a wall, partition or floor that has a ½ hour fire check. Certain cable may also need mechanical protection. FP200 fire resistant cable is now extensively used in the fire alarm sector for both groups and, as it is easily terminated and routed due to its flexibility, it is very much preferred by installers. All cables must be installed in accordance with the good practices advocated in the IEE Wiring Regulations and have a cross-sectional area of not less than 1 mm^2 or, if stranded, of not less than 0.5 mm². They should be routed through low fire risk areas and have a PVC sheath if they run through corrosive damp or underground points. They must also be separated from other cables unless they are in trunking or similar types of channel.

It is worth looking at fire resistant cable FP200 in a little detail because it is being increasingly used within the fire alarm industry. It is supported by a British Standard Certificate of Assessment and as it is also approved by the Loss Prevention Council (LPC) and listed in 'Rules for automatic fire detection and alarm installations for the protection of property' it can be used with confidence. FP200 is available in a wide range of configurations from a 2 core version with an uninsulated circuit protective conductor, up to 19 core with a drain wire. This cable is manageable to work with and is easily formed with the minimum recommended bending radius being 6 D, where D is the overall nominal diameter of the cable. Termination is also straight-forward. In construction FP200 has an annealed aluminium laminate and composite sheath applied longitudinally and folded across the cores to give an overlap, this forms an excellent moisture barrier providing good hoop strength and high frequency screening. The insulation of the conductors is silicone and hence is suitable for continuous operation across the range 0–150°C and is satisfactory for short time operation at temperatures up to 200°C. It is also suitable for surface wiring or direct burial in plaster for both indoor and outdoor installations in satisfactorily protected environments. Extra protection of the cable is only necessary where mechanical hazards exist, when it is buried in floor screed cable ducts or when capping should be used in the installation.

The terminating of the cable is achieved by scoring around the sheath with a knife without cutting fully through the aluminium tape. Then by gently flexing at the scored point the sheath will yield and can be pulled away by twisting gently to follow the lay of the cores. Ferrules may then be slid over the cores and fitted over the end of the sheath to prevent electrical faults occurring from the insulated conductors to the earthed screen.

As no special glanding is required for this cable, nylon stuffing glands can be used, and a shroud then applied over the whole assembly to enhance the termination.

The cable may be required to maintain circuit integrity in a fire, it is therefore important that the cable clip used to fix and support the cable can also provide resistance to that fire. Metal single hole fixing clips with a plastic coating are advocated for this purpose. In an accessible position the spacing of these clips should be at least 250 mm horizontally or 400 mm vertically for cables of up to 9 mm overall diameter, then progressing slightly up to 400 mm and 550 mm respectively for larger diameter cables.

In practice we now find that many fire alarms are wired in FP200 or in one of the many new generation, equivalent type, fire-proof cables for the sounder circuits, with the call points and automatic detectors being cabled in PVC sheathed wiring with mechanical protection then being afforded. This is a move away from the traditional method of using mineral insulated copper sheathed cable with its more complex terminations comprising two subassemblies, each performing a different function. The first is the seal which excludes moisture from the cable at its fixing. The second is the cable anchoring.

It must be stressed at this point that when planning and installing the cabling it is essential that these are routed so that any detector or sounder is incorporated in line. Spurs must not be used under any circumstances as the spur wiring cannot be monitored correctly.

Although, as previously said, call points and automatic detectors can be cabled on the same circuit, it may be advisable for speed of identification that they are on separate zones. Equally, if a detector head is removed, the wiring must be such that the call points are not inhibited. With amendments to BS 5839 in these circumstances the end of line device resistor is replaced with an active end of line device together with the continuity diodes. An option is always to use a control panel with extra zones or to adopt a single loop addressable system.

By means of an addressable fire detection system a large number of devices can be linked by a single screened twin-wire loop circuit which often extend up to 2 km. These tend to use plug-in address modules that are inserted into the appropriately numbered addressable detector base or call point, avoiding the risk of address duplication. In the event of a fire, the address number of the device in alarm is indicated on the control panel display. Should the fire incident spread, subsequent addresses in alarm are stored for future interrogation. Should a fault occur on the address loop, similarly comprehensive information, such as fault condition indication, fault address, subsequent addresses in fault and the total number of faulty detectors, is displayed.

Having completed the cabling, and before any detectors, sounders or control equipment of any kind are connected, the cables can be Meggered to prove no faults exist.

Once the wiring has been completed and the system finalized the sounders must be tested and then tested again at weekly intervals to ensure correct operation. This is a requirement of BS 5839. Further weekly tests should involve the operating of a manual call point or smoke detector with a different detector being verified each week.

Battery connections should be checked and all the operations entered in a log. Although the panel should be checked daily for normal operation indication it should also be checked quarterly for all functions by simulating a fault condition. A visual check should also be undertaken to ensure that no structural alterations have been made that could influence the siting of detectors and other trigger devices. Once a year at least every detector's operation should be verified *in situ* and all cabling should be checked for security and damage. In addition to keeping the system in good operating condition, all these checks can go a long way towards preventing false activations.

4.4 False alarms

In order to reduce the incidence of false alarms, certain precautions must be taken by the installer and be understood by the user of the premises. Common causes of false fire alarms include mechanical and electrical disturbances arising from vibration, impact and inadequate servicing. Ambient conditions including high air turbulence or heat, smoke and flame from a process are obviously troublesome conditions.

Heat detectors can be influenced by abnormal increases in temperature caused by heating equipment, industrial processes and sunshine. Where direct sunlight can occur a shade should be introduced. Detectors with relatively higher temperature settings should be employed where space heating can cause a problem.

When smoke detectors are used attention should be drawn to any processes that can produce smoke, fumes or dust, or fibres, and also to steam and condensation produced by the environment.

Unfortunately factors such as environmental contamination, line faults, sudden rises in temperature and faulty trigger devices can always cause false fire alarms. Such occurrences are extremely counter-productive and furthermore are dangerous. Major new developments in fire detection systems are taking place and the majority of false alarm causes are now being alleviated.

In most cases, with a view to eliminating nuisance alarms, fully or partial complex information processing systems have been developed. In these systems detectors do not of themselves determine if an alarm threshold has been exceeded but it is at the central control panel, to which the detectors have sent their data, that the decision is made. By this method a warning is gained if a detector's sensitivity is drifting.

Using 'intelligent' addressable analogue transmitting, systems can provide a wide variety of information. Each sensor may be addressed in turn by the control panel and a microprocessor based system can analyse the specific sensor data received and then provide a greatly improved indication of the sensor state. Drifts, transients and even factors such as insect strikes can then be recognized. A centralized analysis of sensor trends for accurate and reliable fire or fault warnings becomes apparent. This enables a system to continue operating correctly even when multiple faults become known.

Another way to check the sensitivity of a detector is by using special decentralized system detectors in conjunction with a conventional panel. By this method a whole system can be checked at the panel in a short space of time by pressing a series of zone switches in sequence. If the detector is outside of its acceptable range, a signal indication will be given by the detector and recognized at the panel, although it must be appreciated that detector sensitivity will vary with changes in temperature, humidity and pressure. By this method detectors that are drifting towards constant alarm or are losing their detection properties can be identified. Contaminated detectors can then be cleaned or replaced depending on their condition.

Automatic detectors are much more complex than manual units and it is well known that manual break-glass call points are extremely reliable. In practice they do not normally cause any false alarm condition, therefore their siting is only determined by the following:

- to be located on exit routes and on floor landings of staircases plus all exits to the open air;
- in general, to be at a height of 1.4 m from ground level;
- to be easily accessible, well illuminated and at a position free from any obstructions.

It is advisable that the method of operating all manual call points in a building is identical and that a person need not travel more than 30 m to be able to signal an alarm condition.

There are in fact some self-contained manual call points with integral sounders. Indeed with changes in Building Regulations there is an increase in demand for smoke detectors with integral sounders which can be connected directly to the 240 V ac mains supply. These are called 'single point' and they may also have an inbuilt standby battery. They can be linked with like devices so they feature LED indicators, because any device that triggers will activate the sounders on others. The LED will identify which unit caused the alarm. There is little else to do with this detector type other than to press its test button periodically. However with the British Standard system, regular servicing is needed and maintenance contracts are offered by competent organizations. A full itemized report on every part of the installation should be obtained at least annually. The system inspection is summarized in the following sections.

Daily inspection

● Check the mains LED is lit.
● Check that no other LEDs are lit or that any sounders are operating.

Weekly test

● Operate a call point or sensor to test the alarm.
● Check that all alarm sounders operate.
● Reset the system.
● Each week test a different zone (if applicable). Also use a different call point or sensor for each test, so that all call points and sensors are tested in rotation. A plan of the building detailing the location of all detectors and call points should be provided.
● Check all call points and detectors and verify that none is obstructed in any way.

Quarterly test

● Check all previous log book entries and verify that remedial action has been taken.
● Visually inspect the battery and its connections.
● Operate a call point or detector in each zone to test the fire alarm as in the weekly test.
● Disconnect the mains supply and check that the battery is capable of supplying the sounders with adequate power.

Annual test

● As for the weekly and quarterly tests but check every sensor, call point, sounder and any auxiliary equipment for correct operation.

Every 2–3 years

● Clean the smoke detectors to ensure correct operation and freedom from false alarms. Special equipment is available for this purpose.

Every 4 years

• Replace the standby batteries

Going beyond the requirements of BS 5839 there is now a good selection of battery driven smoke detectors particularly for domestic use and their value cannot be overstated. One can also purchase combined PIR/smoke detectors and these gain a mention in Section 5.2 on stand-alone alarms.

4.5 Ac mains detectors

Changes in Building Regulations have led to a substantial increase in demand for smoke detectors which can be connected directly to the mains supply without the need for a control panel. We do not specifically class these detectors as stand-alone units because they are more aimed at the professional installer market, as mentioned in section 4.4. Both optical and ionization versions can be sourced, although we can say that today optical smoke detectors are more widely used than the latter due to the growing use of flame-retardant materials in building construction, decoration and furnishings. These may also have a 9 V battery back-up, usually of zinc carbon with the mains connection itself via a simple plug-in receptacle located at the back of the detector. The battery can be easily replaced without the need to remove the detector from its fixing.

Options include low battery warning signals with a tamper-proof battery case lid and a full function test switch to prove the sounder operation. Interconnection can be employed using triple conductor mains cable with two cores supplying power and the third carrying a signal between the interconnected devices. In the event that a device goes into alarm the sounders on all units will be activated and the LED on the unit detecting smoke will normally provide indication by going out. Under normal conditions the LED will indicate mains presence.

A relay base can often be used to provide an output to ancillary equipment. The smoke detector will automatically reset and need only be on a circuit that cannot be switched off under normal circumstances and is identified as to its duty.

5 Miscellaneous security methods

In the domestic sector persons who wish to install a comprehensive form of electronic security or call system are often prevented from doing so by cost. In these cases an alternative, less expensive, method may be worth considering – many diverse forms exist. These range from ultrasound devices that react to noise and can trigger lights, to devices that will activate a tape recording of a barking dog. However, it will be found that some of the most common security units in current use are run in conjunction with lighting.

5.1 Lighting

Programmable security light switch

These devices may be installed in 1- or 2-way lighting circuits or in circuits that have an intermediate switch. They do not replace the intermediate switch itself but replace one of the 2-way switches. The programmable security light switch automatically controls lights that are normally operated by wall switches.

These units switch lights on at the zero point in the mains cycle and help extend lamp life by avoiding heavy surge currents. In the programme cycle there is a built-in variability of on/off times, hence lights in a house can be set to come on at quite random times. This can deter any potential intruder if lights in a house are being turned on and off to no apparent sequence. These security light switches also feature manual switching to override the programmable cycle when normal duty is required.

Plug-in timers

Plug-in timers are available in many forms ranging from standard electro-mechanical types to electronic slimline units. In practice they all plug into a mains socket outlet and they provide an interface between the socket and a device such as a standard lamp or radio, etc. They are programmable with a manual override.

Electronic models feature battery cells but are normally powered from the mains when the timer is plugged in; program protection is provided when the timer is unplugged or the mains fails.

In the same way as programmable security light switches, they can switch indoor or outside lights on and off at random or can provide illumination at a specific time throughout every 24-hour period. An override enables them to be permanently on.

Photocell switches

For exterior security lighting applications photocell switches of splashproof construction are useful. These devices automatically switch lamps on as light levels fall below a predetermined level. When the level of light is restored then the lamps will be extinguished.

All these light control units should be designed to conform with governing British Standards since they are effectively mains operated and used in the domestic environment. Governing Standards are BS 800 which specifies the limits of radio interference, BS 1363 which refers to 13 A plugs, socket outlet and boxes and BS 3955 which covers the safety requirements and tests for electrical controls.

Photocells do not need any particular orientation but should be faced away from artificial light sources although the circuity does include a built-in time delay to prevent any sporadic unwanted switching as the result of periodic light sources such as car headlights. Often homeowners only want to use generally available bulkhead lights, but recognize the inconvenience of only being able to activate these from the interior of the house. When the homeowner leaves the house the lights must remain on in preparation for his or her return. This can be inconvenient and costly, since the lights are illuminated for periods when it is not necessary. If one does not wish to go to automatic security lighting or photocell switches, then an inexpensive outside splashproof domestic light switch could be utilized. These are purchased in the form of a robust metallic weatherproof housing which can contain an architrave switch electrically rated for the intended duty. Figure 5.1 shows a useful application method where a switch is fitted to control lights installed for safety along a dark passage.

The three lights shown may be switched on from inside the house before exiting through the door and then turned off from the weatherproof switch when leaving the dark passageway. The opposite sequence is followed on returning to the premises.

This is a very efficient security and safety lighting method and leads us on to look at how basic stand-alone or self-contained devices can be of great assistance and yet not introduce any great complexity.

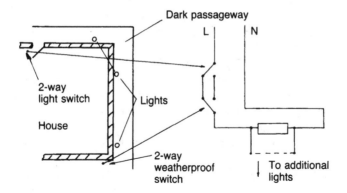

Figure 5.1   *Switch controlled lights*

5.2 Stand-alone alarms

Intruder detectors

In order to protect a whole room or a strategically important area through which an intruder must pass, self-contained security alarms can be considered. These are stand-alone items often found in hallways or the main lounge of domestic properties.

These alarms are normally of a portable construction, battery operated and use passive infra red, dual element, pyro-electric sensor principles. The method of operation is similar to that detailed in earlier chapters of this book.

The systems can be free standing or wall mounted with arming and disarming being affected via a membrane keypad or keyswitch. Entry/exit times permit a delay to turn the system to its on or off mode. Battery life is monitored and indication of low charge is given for a period before exhaustion. Units are also tamper protected so that any movement of it once armed causes an instant alarm. Although stand-alone systems have a built-in siren, additional contacts tend to be provided so that it may be linked to lights or extension sirens.

Purpose designed extension sirens can increase the audibility of stand-alone alarms, which are wired to the main alarm in such a way that they sound at the same instant. Cutting the wires between the master unit and an extension siren will cause the extension siren to sound due to its inbuilt cell.

Although easily transported domestic appliances such as hi-fis, videos, televisions, microwave cookers and computers, etc. can be protected by putting them in the vicinity of stand-alone intruder alarms, they can also be protected by having limpet type seismic detectors fitted to them. These devices prevent articles being carried away. However, there is now a move to fit many valuable articles with integral sirens that are 'held off' electronically provided that the appliance remains connected to the mains. If any disconnection is made an alarm condition is generated. In their simplest form these devices can be installed as an interface in the wiring between the appliance and the mains connection.

This form of alarm generally falls into the class of self-contained systems and would not under normal circumstances have any external sounder mounted on the premises. It would, nevertheless, intimidate any intruder and would under normal circumstances achieve its aim.

Bicycle alarms

Motorbikes, pushbikes, trailers, large lawnmowers, etc. can be protected against theft by the use of high output audible alarm locks. These feature a steel cable that is secured to railings, lampposts, etc. and then inserted into a cable locking socket. A tamper switch and vibration sensor protect against any cutting of the cable or tampering with the lock. Confirmation beeps are emitted by the unit to confirm battery condition and setting when the unit is switched on. Also when the alarm is switched on and the lock is moved in any way, there is a delay of 5 seconds and then the alarm will automatically sound for a period of 15 or so seconds. This cycle of 5 seconds on and 15 seconds off will continue only if the alarm is continually tampered with. Mounting brackets provided with the device are purpose designed for fixing to the specific frame type. These items, which work off 9 V or 12 V batteries, are invaluable because they ensure that a thief cannot work at cutting or forcing any lock without incurring attention from any persons within hearing distance.

Child alert alarms

Child alert alarms are generally light sensitive and designed to warn parents or child-minders that a drawer or cupboard has been opened. The alarm activates a sounder when small hands open a medicine cabinet, a drawer, or enter a desk which is intended to be protected. An on/off switch, which is awkward for youngsters to operate, is employed which, when turned on, locks the cover which surrounds the battery. Once set it allows a measured amount of time for it to be placed into the protected area. Once activated and exposed to light it will trigger an alarm condition and normally stay in alarm until switched off. Once again it is classified as a stand-alone unit but it can be fixed in temporarily by means of double sided sticky pads or tape.

Keypad entry door security alarm

Keypad entry door security alarms are highly versatile and can be easily installed on the inside of a door with a magnet assembly that is mounted on the frame. When the door is closed the magnet is detected by a proximity sensor but when it is opened the sensor recognizes the change and activates an alarm. This can only be switched off by entering the correct code on the main unit's keypad. These devices tend to have other features such as 'chime' to announce the entry of visitors, an instant alarm should the door be opened when the device is set, and delay set for an authorized user to enter and leave by the protected door. This operates in the same way as the entry/exit time delay seen in the well-known intruder alarm control panel. The sounder output can also be used to drive extension sirens if required.

In earlier sections of the book we looked at access control, in which persons are allowed access to a given area protected by an electric lock or electro-magnet that is switched when a recognized entry procedure is followed. There is on the market nowadays an expanding range of easy to install electronic keypads for both internal and external use and these, when integrated with security power supplies, act in the same way as a keypad entry door security alarm with access control. As a progression from this, there are also rapid developments in door security methods using infra red techniques.

Infra red door security kit

These are designed to provide hand held remote door entry at the touch of a button without compromising security. They can be used on most internal and sheltered external doors where high security and simple to use access control technology is wanted. These products, which are security protected, use an infra red key which incorporates an operating button.

The key is pointed towards a receiver which is sited adjacent to the door being protected and can be some 3 m or so away. The receiver then sends a signal to a decoder, which is itself mounted on the secure side of the door, and this checks the signal to ensure that it contains the security code. If so the decoder pulses the electric strike to unlock the door for a preset selected period of time. The system is, as usual, powered by low voltage via a transformer wired to the mains.

The security of these infra red keys is high and cannot possibly be influenced by remotely operated car alarms or television remote controls which use similar techniques. Additional keys can always be supplied on demand by the kit manufacturer.

Simple locking methods performed electrically from a remote point to allow or deny access can be useful in places where late night attendants operate. Examples are petrol forecourts and off-licences. These locking methods can allow doors to be left unlocked under normal circumstances but they may be quickly locked if the operator wishes to deny entry to a particular individual. The infra red technique could be used for such applications, or a one-way switch operating an electric strike via a power supply or transformer could also be employed.

The majority of places that could be open to the public late at night will probably have an intruder alarm already, so a personal attack button could also be placed adjacent to the device controlling the door strike. The operator is thus given increasing levels of electronic protection – extending from locking a door, to additionally generating an audible signal to summon help, or perhaps even using remote signalling to a manned central station.

Personal security alarms

In Section 1.5 we looked at personal security devices often called PA (personal attack) pushes and how they are hardwired into a security system. There are also many portable versions which fall into the stand-alone category. First it should be stated that a personal attack alarm is the only deterrent allowed by law. Street crime is expected to increase and those most at risk include people who work early or late shifts, walkers, joggers, women, children and the elderly or anyone who travels alone by foot during unsociable hours. The police and consumer groups recommend the carrying of pocket alarms by these at-risk groups as they draw immediate attention to the victim and attacker providing a deterrent to the criminal whilst summoning help. The units are of rugged construction and need only have their batteries checked periodically to ensure that the alarm will operate when needed. Devices can be easily carried in a handbag or pocket or attached to a belt loop. These are battery powered and can be actuated by pulling a ripcord which detaches a plastic plug from its socket. It stays in alarm until the plug is replaced.

Personal security alarms can be used with straps attached to a handbag and should theft be attempted the strap will be separated from the socket actuating the alarm. This draws attention to the handbag which is connected to a further longer strap. Variant devices include a flashlight which can function as a torch under normal circumstances, using an exclusive switch, but which, if activated by the cord, provides a flashing beam of light to coincide with the audible output.

Smoke detectors

For basic safety, battery powered stand-alone ionization or photo-electric smoke detectors can be installed at little cost. Ionization units provide early warning of developing fires especially of a fast flaming nature. These are self-contained units using a dual element ionization chamber and come complete with a warning siren which also provides low battery indication by bleeping at regular intervals. A test-on enables the homeowner to electronically simulate the presence of smoke. The unit resets itself and deactivates the internal siren once the smoke clears the area. More advanced units have hinged lids to make fixing and battery replacement more easy and others come complete with an integral escape light to assist the escape process.

Photoelectric units using the scattered light principle and with insect resistant screens to prevent false alarms are also available. These items provide early warning of developing fires, especially of the slow smouldering type. Like the ionization version they come complete with a siren but this type can be interlinked, so that if one detector senses smoke then all linked units will sound the alarm. However the low battery signal sound will only occur in the unit needing a new battery. Indeed battery operated smoke detectors of all types are rapidly gaining popularity for use in the domestic sector. In practice they should be installed as close to the centre of the ceiling as is possible and be placed outside every bedroom area and on every level of the home. If bedroom doors are usually in a closed position at night then a detector should be installed in each bedroom as well as in the common hallway between them. It is advised that smoke detectors in multi-level premises be interconnected so that a detector on any level of the protected premises will generate an alarm of sufficient audible output to awaken sleepers in closed bedrooms.

Detectors should also appear in every bedroom where a smoker sleeps and in sleeping areas where portable type heaters and humidifiers can operate during the night. If a bedroom hallway is of a length in excess of 12 m then detectors should be installed at both ends of the hall but they should not be placed closer than 10 cm to a wall or corner. If ceilings are sloped or peaked, detectors are best put some 0.9 m, measured horizontally, from the highest ceiling point.

Mobile homes and caravans of modern construction should have detectors placed in the same way as in the domestic home, whereas older mobile homes and caravans with little insulation are best protected by detectors placed on inside walls only.

Areas where placing smoke detectors should be avoided depend very much on the detector type and manufacturers' data sheets should be consulted. Basically they should not be put where combustion particles are normally present such as in kitchens, in garages, near furnaces or gas space heaters, etc. Equally they should not appear in the air streams passing by kitchens.

Smoke detectors must be tested weekly using the test switch. The battery should also be checked for efficient function and dust should regularly be vacuumed off the detector and the sensing chambers kept in a clean condition.

Combined PIR and smoke detector

Although not a stand-alone unit as such the space saving combined PIR and smoke detector can usefully be discussed at this point. It is suitable for use in any intruder alarm system with a 12 V detection system. The PIR function fulfils its role as an intruder detector and the smoke detector, normally an ionization type, has separate terminals for use with a control panel which has fire and intruder outputs. Eight core cable is used to wire the unit, two for the power connections, two each for the smoke detector and PIR, and two for the anti-tamper function.

Carbon monoxide detectors

Carbon monoxide (CO) is the most common cause of death by accidental poisoning in the home in the UK. It is a gas that cannot be seen or tasted and cannot be detected by any human senses as it has no smell. In practice it can leak from any appliance using gas, coal or oil and is produced when the appliance is either not working correctly or is badly vented. Fatalities occur in concentrations in air as low as 3–400 ppm (parts per million).

It is possible to purchase carbon monoxide detectors that can be linked to an existing security system or to buy stand-alone devices that are mains powered with a back-up battery. There are essentially three different types: 1, the semiconductor version measures levels of CO at different temperatures; 2, the gel cell is of a porous material which darkens when CO is present; and 3, the electrochemical cell works by electrochemically oxidizing the CO to form carbon dioxide and is generally considered the most reliable.

The installation of these detectors is straightforward. They should be sited in or near every room that has a cooking or heating appliance and be positioned at least five feet above floor level. They are always best placed higher rather than lower but should be at least six feet from any appliance being monitored.

5.3 Wireless intruder alarms

Wireless intruder alarms are also often called radio systems and work by linking detection devices to radio transmitters which communicate with a central control unit. These systems are particularly easy to install as they require little wiring but at this stage in time they are more expensive to purchase in component form than are their hardwired counterparts. They are economical in terms of installation costs and are becoming increasingly popular, indeed they hold great potential for the future. Wireless signalling systems consist of a combination of radio receivers and transmitters used in conjunction with intruder detectors that are battery powered. A selected system code ensures that the central control unit receiver can only respond to transmitters set to the same code. Accidental or deliberate interference from alien transmissions is therefore negated. The effect of crosstalk between adjacent systems and stray radiation is also ignored. The system itself is supervised to provide tamper, low battery and 'jamming' conditions.

The system controller provides a specified number of zones and usually features an integral siren. It needs to be connected to a 240 V ac mains supply but is equipped with a back-up battery source to give power loss protection. Being supervised the sensors report to the controller so that a visual status of the system's transmitters is always provided. Movement detectors used in wireless alarms use passive infra red detection principles and transmit their alarm signals back to the system controller. These PIR detectors are battery powered, compact and completely self-contained since the transmitter is contained within the detector housing. Such units are fully tamper protected, wall mounted and generally have detection ranges of 10–12 m and employ a recognized walk test facility. Advanced movement detectors also incorporate pairs of normally open and normally closed contacts for optional additional use with magnetic switch contacts, pressure mats or vibration detectors. Transmitters can be activated by external magnets fitted to doors or windows. In this case the transmitters take the form of a stylish housing and come complete with a reed switch. Inputs are also provided for connection to any detector which has little or no power. In this context we class flush or surface mounted magnetic contacts or microswitches plus pressure sensitive mats. A single contact transmitter may be connected to more than one set of switches. It usually features an LED, is fully tamperproof and can signal the switch opening. If there are aesthetic reasons for not wanting to mount the transmitter in the vicinity of the door or window, then it can always be hidden under a curtain to protect adjoining doors or windows. Wires may then be run from the detection device to the transmitter only.

Emergency pocket sized hand held transmitters are effective anywhere within signalling range of the system controller whether the alarm is in a day or set condition. These transmitters perform the function of portable panic alarms. They are also available in a dual format where they perform as a panic alarm and remote controller. Whilst the panic button will activate an output at the receiver at any time (useful for mobile security guards) the remote on/off button can cause a trigger capable of setting or unsetting the central control panel.

Wireless interior sirens that plug into an electric socket and receive line carrier signals from the central control unit can be purchased. External weatherproof sirens with or without a strobe light are activated by radio signals from the controller. These are self-activating if they are tampered with or if any attempt is made to remove them from the wall. Their power is derived from the 240 V ac mains supply and is complete with a rechargeable battery to give power loss protection in the event of mains failure.

BS 6799 'Code of Practice for Wire-Free Intruder Alarm Systems' contains criteria for the construction, installation and operation of intruder alarm systems in buildings using wire-free links between components, for example radio or ultrasound.

This British Standard covers systems that consist of one or more transmitter units associated with a detector or detectors and a common receiver unit. The code of practice identifies five levels of system performance in an ascending order of sophistication, known as Classes I to V. These are clearly defined in the Standard: a product which comes into Class I is seen as being at a basic level, Class II provides an identification at the receiver that the transmitter has operated. Class III is as Class II but with the communication channel in use monitored, so that any continuous blocking or interfering signals that are present for more than 30 seconds and that could prevent the reception of legitimate signals are detected and a fault indicator is generated at the receiver. Indeed it is to Class III that most domestic systems are designed to comply since the highest level Class V calls for sophisticated auto-reporting at specified intervals and alarm condition generation.

The use of radio frequency transmission by wire-free intruder alarm systems is controlled by the Department of Trade and Industry,

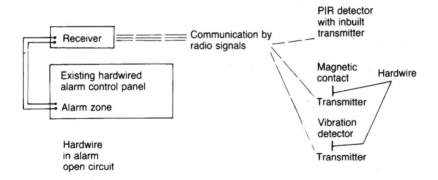

Figure 5.2 *Wireless configuration*

Radio Regulatory Division, since the systems have to comply with the appropriate regulations. It will be found that many of these supervised radio alarm systems use a microprocessor based alarm control panel that has similar functions to an already established hardwired counterpart well known within the industry. It also enables someone familiar with a specific hardwired unit to be able to diversify, in fact radio panels do also sometimes provide some hardwired loops.

Wireless detection devices are not unlike their hardwired equivalents, and the methods of employing and positioning them are identical. The radio components are available as a range of units fully compatible with virtually every hardwired control panel on the market. These provide the ultimate in convenience and performance when wiring is not an option. It is therefore possible to purchase radio detectors, transmitters and receivers which are linked to an existing hardwired control panel and system via the integral contacts in the receiver (Figure 5.2).

If we look beyond the addition of radio components to an existing hardwired system and towards a fully programmable professional radio security system then we should consider any limitations this may have on the positioning of the main control unit.

Wherever possible the main control unit should be installed as near as possible to the centre of the group of detectors/transmitters to achieve approximately equal signal strengths. It should also be at least 1 m from large metal objects such as radiators and not be fixed underground or in a basement. Equally it should not be in close proximity to electrical switchgear, televisions, computers or similar sources of radio interference. The ideal height is some 1.5 m from the floor and at eye level. If the premises are of reinforced concrete or steel construction then a field strength meter will have to be employed to ensure that the signal strength is adequate – this will be quoted in the installation manual. The only wiring needed will be the installation of an unswitched fused spur adjacent to the control panel for the mains connection. An external buzzer or extension speaker, if added, must be hardwired, otherwise the only wiring will be to the external sounder/strobe via an SAB unit. This cabling is essentially the same as that shown in Figure 1. 7. For full flexibility the professional unit also caters for the use of a wireless smoke detector which is active 24 hours and initiates a full alarm if activated.

The reader will clearly note that the advantages of the wireless system lie in its ease of installation due to only a minimal amount of cabling needing to be employed. However, it must also be understood that since the detection devices and transmitters are battery powered an ongoing programme to change and recharge these cells must be initiated, although supervision will advise of low battery condition. We know that professional wireless systems are more expensive to purchase as components than the equivalent hardwired systems and, although savings can be made in the lower cost of installing a radio system, the advantage of this must be balanced against the benefits of the hardwired system in the long term.

We can say that any alarm system can only be a secure as the communication link between the detection devices and control panel, and we are now finding that radio signalling, with its immunity to line interference and attacks, is a great step towards secure signalling. Paknet is a radio signalling service launched on to the security industry to address line attacks and is now offered by many alarm monitoring stations. It is discussed further in Section 5.5 which deals with 24-hour central stations.

5.4 Vehicle and special purpose protection

Vehicle protection and alarms cover a wide range, since many different systems are available and the application of them depends very much on the level of protection required for the vehicle.

The basic system consists of an ignition inhibitor switch which is concealed in the vehicle and prevents the engine from starting. This is achieved by installing a simple on/off two terminal switch in the low voltage wiring between the ignition switch and coil or by using the switch to open the wiring to the fuel or diesel pump making starting or running impossible.

Burglar alarm control modules wired to the existing door switches, and needing a special key to turn them off, can cause the vehicle's horn to sound and the headlights to flash for a prescribed period if unauthorized entry to the vehicle is made. Alternatively a special siren can be fitted that is independent of the car horn. Ignition inhibition can also be employed with these systems.

When the contents of the car need to be protected and the owner fears that entry could be made to the vehicle by breaking a pane of glass, then ultrasonic motion detectors can be fitted and these are available in many different forms. The system is delayed if legitimate entry to the vehicle is made, otherwise the ultrasonic motion detectors trigger audible and/or visual alarms such as the horn and flashing of indicators or headlights. Pendulum effect switches, forming part of the system, detect movement of the vehicle or a door opening. Careful adjustment is needed to ensure the correct sensitivity so that a slight movement of the vehicle due to the wind or passing traffic does not cause spurious action. Many vehicle alarms employ a combination of the aforementioned detectors whilst for setting and unsetting the alarm they may use infrared or radio signal principles rather than keys.

This setting and unsetting procedure normally flashes the direction indicators with a different number of pulses to confirm whether it is being switched on or off. A warning LED on the dashboard lights to indicate being in the set condition. An example of an alarm that uses a combination of detectors is shown in Figure 5.3.

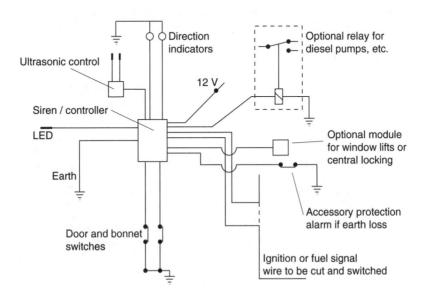

Figure 5.3 *Alarm using a combination of detectors*

The alarm shown in Figure 5.3 gives space protection to the vehicle and also protects the doors and bonnet against forced entry. Interfaces for central door locking and electric window lifts can be added so that on setting the alarm all doors and windows automatically lock. Extra features include engine immobilization where the alarm has been triggered and accessory protection. If any attempt is made to remove mobile phones, radio/audio equipment, fog lights, etc., then the alarm will be activated. An input for a tilt sensor is often incorporated and can detect whether the vehicle is being tilted or lifted, for example if somebody is attempting to steal the wheels. Certain remote setting key fobs have a further switch which can select an immediate alarm and so operate as a personal attack (PA) function.

Vehicle alarms with battery back-up have a small support battery held in a compartment of the siren to give additional security. When armed, if the main power or earth connection is broken to the alarm, then the alarm is triggered by the power derived from the support battery. A key operated switch is usually supplied on the back of the alarm and will effectively bypass all the alarm functions when turned off. This is convenient when the vehicle is being serviced or valeted.

At the present time there is also a range of immobilizers available which operate by means of a jack plug. These scramble electrical and electronic circuits in the car's system making starting and running impossible. However for the reader who does not wish to work with any complex wiring there is a good range of alarms available which use a voltage drop method and are not difficult to install. These sense an interior light coming on, e.g. through the opening of a door, boot, bonnet, glove compartment, etc. (Figure 5.4 shows an example of this).

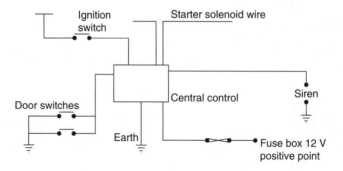

Figure 5.4 *Voltage drop/current sensing alarm*

These systems are installed by running one wire from the central control to one of the existing door switch wires and connecting to a point on that wire. One only needs to ensure that all the doors, when opened, do bring on the interior light to use the voltage drop method since it is the power derived by the interior light that is recognized. Further wires are run to an earth point and another to a 12 V positive point in the fuse box. The siren is connected in a secure position to which it is earthed and then its supply wire is run back to the central control. A wire is next connected to the terminal on the vehicle's ignition switch which has 12 volts present in the starting and run positions. The object of this wire is to ensure that the system cannot arm when the vehicle is running. Immobilization of the starter motor is effected by running the final two wires to the starter solenoid wire which is cut and the two wires are connected one to each of the cut ends. These alarms are generally turned on and off from the outside of the vehicle by a remote setting fob.

There are keyless delay alarms available which use the voltage drop method and only need the connection of some three wires under the bonnet. They work automatically from the ignition – when it is turned off an exit cycle starts and the owner leaves the vehicle. On re-entering the vehicle an entry cycle starts giving the owner a few seconds to turn on the ignition, which switches off the alarm.

Many of the products discussed feature an extra wire that can be connected to electric cooling fans of the type which still operate for a period after the ignition is switched off. This extra wire is used to prevent the current sensing on that circuit triggering the alarm.

Portable car alarms

As with other sectors of the security industry stand-alone or portable vehicle alarms are also available. For those who do not wish to do any permanent wiring a portable car alarm with remote control could be considered as these can be swapped between vehicles. The unit is mounted internally, close to the dashboard; the main control derives its power from a standard cigarette lighter socket via a coiled lead. These portable alarms have their own back-up battery which will keep sounding the built-in siren if it is disconnected from the lighter socket. LEDs indicate the state of the back-up battery and mains supply. The alarm is triggered by a shock to the vehicle or by a voltage drop in the vehicle's electrical supply. The latter can be configured to delay current sensing for a period after arming – necessary if the vehicle has an electric cooling fan.

It has long been recognized that no device can offer complete protection against theft but that even the simplest of alarms or immobilizers is a great deterrent and a huge irritation to those attempting to overcome them. Even the most inexpensive and basic of those considered in this chapter are very much worthwhile in their own right.

5.5 24-hour central stations

Many of the intruder and fire alarm control panels detailed in the appropriate chapters in this book are capable of performing remote signalling. That is they may provide a signal over the telephone network to a remote station that is permanently manned, in addition to providing audible and/or visual warning at the location of the generated alarm. This has always been seen as necessary for intruder alarms installed in premises that are in an isolated position where a local alarm is not readily detected. Rural premises fall into this category or factory sites that have no evening watch or night working.

Originally 24-hour manned central stations were intended to monitor intruder and fire alarm signals so that response could be made on immediate receipt of the telephone message. The signalling of alarms to the centre is often over direct lines using the traditional multiplexed working technique. Alternatively by using a printed circuit board digital communicator, integrated with the control panel, an automatic dial-up method can be employed from the subscriber's premises over the PSTN (public switched telephone network) to the monitoring station. Using the public telephone network was always viewed as easy to install and more economical than employing a direct line. Using digital communicators calls for a special ex-directory exclusive line for the alarm system. This is used to dial the centre directly and report a given message. In case the message did not get through at the first attempt, further separate programmed attempts would be effected to ensure that communication was made. Alternatively other multi-digit numbers could be dialled if no response from the first selected number could be gained. Although many automatic diallers connected to intruder or fire control panels dialled 999 direct if an alarm at the panel was made, a more practised and recommended method of reporting alarms to the required service now exists. This is usually over a telephone line from the communication equipment to a permanently manned station operated by an intruder alarm company. These companies are in communication with the relevant command headquarters for the police and fire brigades and the services can respond, providing a unique registration number can be quoted. Often 24-hour central stations have security patrols of their own which can be sent out in answer to alarm calls. It will usually be found that when communication equipment is fitted to intruder alarms, audible signals are delayed for a prescribed period once the control panel has signalled an alarm. This is to give emergency services the chance to apprehend offenders since they may not realize that the system has triggered.

In fact the role of the 24-hour central station is now being expanded into further communication areas, including the handling of CCTV and slow scan surveillance. Although the central station can retain intruder and fire alarm monitoring as its base function, expanded services can be offered related to safety within commerce and industry. For instance monitoring of plant breakdown, lighting and temperature controls plus the metering of units of electricity, water or other unit types can be provided. Critical conditions of temperature and humidity can also be catered for.

Distress signals or medical alarm monitoring in the care sector could also be effected plus telecontact methods, in which named persons could be contacted through data transmission once a need for their services was received at the station. Using integrated audio channels combined with video systems, such methods can help an operator to obtain accurate information following an alarm condition since intruders can be heard moving around a protected area. This could also help to distinguish between genuine and false alarms. Also it is hoped that in the long term cellular ratio network signalling could be performed where no telephone lines are currently available.

It will now be clear to the reader that all communicating systems in Britain make some use of British Telecom's (or competing companies') services to transmit their signals to central stations or emergency services. This is via the public telephone network or down a direct line. However, it is known that there is a need for continuous verification that secure communications are possible, i.e. line monitoring to report faults that could occur. British Telecom's Communicating Alarm Response Equipment RedCARE enables central stations to receive appropriate signals from their installation in protected areas and it provides a high level of security. RedCARE offers continuous line monitoring without interference, and is done over ordinary telephone lines. It monitors the line, alerting the central station if an alarm is sent or if the line is cut or becomes faulted. In essence British Telecom accepts the responsibility for relaying the alarm signal whilst receipt of signals at the central station and customer contact remain the responsibility of the alarm company.

To ensure that the appropriate response is made quickly the RedCARE transmitter, called Premises Alarm Communication Equipment (PACE), offers different channels to differentiate between differ-

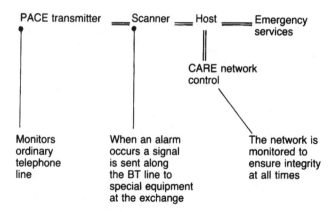

Figure 5.5   *PACE transmitter. Red CARE technique*

ent signals such as intruder, fire, or personal attack as arranged with the alarm company (Figure 5.5).

We have previously looked at wireless or radio intruder alarms and nurse call systems so we should now consider Paknet which is an alarm signalling service launched on to the security industry and now offered by many central stations. Paknet was timed to meet the challenges and requirements of a secure transmission medium which had mainly been dominated over the years by digital communicators on unmonitored lines.

We accept that the alarm system can only ever be as secure as the communications link to the central station and it was believed that using radio signalling with its immunity to line cutting attacks could be a major step to secure communication. In practice this is achieved by high speed digital radio transmission through a national network of radio base stations.

The installation involves the fitting of a special interface device and a small antenna. The interface device links direct to the alarm panel and contains the radio pad which sends and receives the alarm information over the radio data network. On alarm activation, the interface instructs the radio pad to send a short message via the nearest Paknet base station through to the central station which returns an acknowledgement code and also contacts the emergency services.

Paknet is flexible in that it can be adopted with new or existing alarms and a service availability indicator is displayed on the equipment as an assurance.

Statistics have always shown that, from the point of view of commercial risk, the telephone line is liable to be attacked and cut. A method of dual signalling of sequential confirmation using Paknet backed up by a digital dialler makes this risk acceptable and balances it by giving a secure combination. Paknet added to an existing digital communicator system can monitor the telephone line for signal failure or line cuts whilst the digital communicator monitors the Paknet link. Each service essentially monitor the other. The Paknet interface relays to the central station of the line being cut.

The role of the central station or alarm receiving centre (ARC) as it is often called must be both comprehensive and diversive to embrace all modern forms of alarm signalling together with traditional guard response plus paging with message handling. A commitment also exists to filter out false alarms being conveyed to the emergency services.

One problem area with intruder alarms has always been operator error at the opening and closing periods of a premises. It is now ACPO policy that all monitored sites must operate either open and close or alarm and abort signalling at the opening and closing periods of a premises. This is intended to filter out false alarms calls due to user error at these opening and closing times. Two levels of operation exist. Open and close logging is the first, where the signals are logged but only acted upon in the event of an alarm activation. The second form is open and close monitoring in which the central station is automatically alerted if a site is not opened or closed at its predetermined time. This avoids a situation where a site is vacated without the alarm being set. It also enables the customer to receive a report detailing the open and close signals. It is the communication equipment connected to the control panel in the protected premises which has the outputs to provide these signals, which are then analysed at the central station.

In order to be able to perform these functions the alarm receiving centre must have a commitment to quality with approval by the governing body NACOSS to BS 5979 and ISO 9000. Its construction and operation must be to BS EN ISO 9000 and a category of BS 5979 with further approval by the British Security Industry Association (BSIA), the Association of British Insurers (ABI) and the Loss Prevention Council (LPC). The main standard, BS 5979 'Code of Practice for Remote Centres for Alarm Systems', details the planning, construction and facilities of manned and unmanned centres for intruder, fire and social alarms and other monitoring services.

Alarm systems only provide security if a reaction by someone occurs to combat the threat. Often alarms are ignored, hence only security monitoring can ensure a guaranteed response from a designated person and give full electronic surveillance 24 hours a day, 365 days a year. The benefits of a central station are immense and, if affordable, give the client total peace of mind especially if alarms are installed by companies with expertise. These companies may also promote alarm audio verification kits. These interface with existing alarm monitoring panels to allow false alarms to be filtered out and are compatible with the majority of control panels. In the event of an alarm, once the control panel has achieved its communication with the central station it allows the station to call back in order to hear what is actually happening at the protected premises. With several audio channels being used, connected to hidden microphones, large physical areas can be covered and the operator at the station can switch between recorded and live monitoring. Alarm events and information is stored within the network.

In summary it will be well seen that we can all afford, by one means or another, to improve, by an electronic technique, the security of our premises and our well being. The extent to which we bring this electronic protection and security into our lives must be determined by our own financial standing and judgement. It is genuinely hoped that this book has shown the reader a method that can be adopted for each individual case. Indeed, throughout the book we have covered many diverse systems and within these many different cable forms are used. The final chapter is dedicated to the protection of wiring systems and also the electrical safety of such systems when they are mains connected. We also extend our thoughts to batteries, plus the power supplies and control equipment in which they may be found.

6 Wiring systems

The reader will be aware that, although we have also studied systems that are powered solely by batteries, most security systems are permanently connected to the mains supply, with batteries only being a secondary supply or for standby purposes. There are many precautions to be made prior to making the mains supply connection and these are covered in this chapter. Having gained a knowledge of the requirements for making the mains connection we can then consider power supplies and batteries.

There are also general methods used to protect all cables for the different voltages from both mechanical and electrical sources and to ensure that they are adequately supported.

6.1 Inspection and testing of the mains supply

The Electricity at Work Regulations require that people who conduct these tests are competent under the terms of the Health and Safety at Work Act, so the person who carries out these tests must be able to provide evidence with respect to the required competences.

The requirements which we are interested in extend to earth continuity, polarity, insulation and earth loop impedance. After completion of any new wiring installation or alteration to an existing system, the work must be tested and inspected to ascertain that there are no defects and that all necessary conditions have been satisfied. However carefully an installation has been completed it is always possible for faults to occur at a later stage by damage being made to cables by nails or other building materials, or by connections being broken or by defective apparatus being installed. A full test of the mains connection by the security installer should be carried out and the results carefully recorded. In the UK the procedures for inspection and testing on completion of an installation are covered in Part 7 of the IEE Regulations, which includes a checklist for visual inspection. This visual test covers the mechanical protection of the cable and housings. This is to check for damage to insulation or for broken or incorrect connections or for breaks or cracks in junction boxes. It is also necessary to ensure that the cable route is satisfactory and the wiring is adequately supported and the correct capacity of wiring has been used. The cable itself must be mechanically and electrically protected against fault conditions.

Having satisfied the visual examination we can then look at the earth continuity, polarity, insulation and earth loop impedance. In practice the requirements of the IEE Wiring Regulations extend well beyond these tests but the security system or mains emergency or lighting, for security purposes, will only make up a small portion of the full electrical system of the premises. For this reason, providing all of the other appropriate tests have been conducted, we can be assured that the continuity of protective conductors and bonding of earth connections for safety have been guaranteed.

Earth continuity

This test is to ensure that there is continuity to the main earthing terminal at each outlet on the circuit. It is performed by disconnecting the supply and testing between the phase or neutral conductors and the protective conductor at each circuit outlet using a hand operator or other portable device. There are no specific values invoked but it is generally done in conjunction with the earth loop impedance test.

Polarity

This test is necessary to confirm correct polarity in that it checks that the phase and neutral conductors are not reversed and that the neutral conductor is not fused. There is no specific test type in the Regulations but a bell and battery set can be used with a long roving lead taken back to the phase live lead connected to the load side of the meter and with a short lead taken to the apparatus installation point. If the polarity has not been reversed in the wiring run the bell will ring when the roving and short leads are connected to the same conductor.

Insulation

This involves a test for resistance between the pole or the phase terminal and the earth terminal with a dc voltage approximately twice the normal rms value of the working voltage. The insulation resistance to earth between the linked phase and neutral connections should not be less than 0.5 megohms. Often an installation must be sectionalized if a faulty circuit is found because the insulation resistance may be generalized in the system. For this reason it is always advised that the security installer has a unique cable run back to the consumer unit so that it cannot be governed by other cabling.

Earth loop impedance

This, in practice, is the earth fault current loop. A test is done from the phase conductor to the protective earth conductor, which is the path that would be taken by any fault current. This would be done as a test for the full electrical system with the permissible values of ELFI taken from charts in the IEE Wiring Regulations. It is important that the security installer and reader are aware of these conditions otherwise they may well make connections to a system that of itself is not adequate at source.

Usual practice is then to connect the control equipment/power supplies via a spur unit within arm's reach of the equipment so that the equipment can be isolated before the control panel or such is opened. The spur will be fused as appropriate and 3 A is typical for intruder alarms with at least 0.75 mm cable providing the load to the control panel. There is no statutory call for the spur to be fed direct from the consumer unit but it is important to ensure that it is not possible for someone to inadvertently isolate or restore power from another source. For this reason it is best that the supply is direct to the consumer unit but, if this is not possible, any device that could disconnect the supply should be marked to indicate that it also controls the security system. With intruder alarms an unswitched spur is used and this is also the case for fire alarms and most other security control equipment and power supplies, but with some automatic lighting systems switched spurs can be adopted.

The engineer must never work on any electrical apparatus that is live to the mains supply and must power down before modifications are made. This applies when changing printed circuit boards, when the batteries must also be disconnected.

It is important to appreciate how fixed installations and portable apparatus are correctly protected against overload and so that there is no risk of fire in the event of a fault developing. Before we consider this aspect and what the installer meets at the consumer unit or spur we should go back a stage to see how the mains ac supply is actually derived.

Power is generated almost everywhere as alternating current (ac) which means that the current is changing direction continually. In the UK this change of direction occurs 50 times per second and so is classed as 50 Hz (the number of cycles per second). The machinery operating this power is known as a generator and it has three sets of windings in which the current is generated. One end of each winding is connected to a common, or star, point which is termed the neutral.

The other ends of the windings are brought out to the three wires of the phases or supply cables. For identification these are coloured as red, yellow and blue and the currents which are transmitted in each phase have a displacement of 120°.

Power supplies to towns or villages are provided from the power stations via a system of overhead and underground mains and transformers which reduce the voltage in steps from the high transmission voltage to the normal mains voltage at the consumer end. From the substations low voltage overhead and underground mains are taken to the consumers' supply terminals with the ac power being distributed mainly on three phase networks. Thus, on the low voltage side of most local transformers one will find four terminals – the red, yellow and blue phases and the neutral. Between each phase and neutral there is a voltage of 240 V. However, between the phases the voltage is 415 V. For this reason the secondary output voltage of transformers is given as 415/240 V. The supply authorities are required to maintain the voltage at the consumers' supply terminals within plus or minus 6 per cent of the nominal 240 V.

Most domestic consumers are provided with a single-phase supply, unless a high loading is expected. Most industrial premises will have a three-phase supply, as will many commercial buildings.

Power transformers are more efficient if the load on each phase is almost the same, so single-phase services are normally connected to alternate phases and the three-phase consumers are encouraged to balance their loads over the three phases.

The local boards provide underground cable or overhead line services which terminate at a convenient point within the premises. Overhead services are terminated on a bracket high up on a wall of the property and insulated leads are taken through the wall to the meter position. Underground services tend to be brought through floor level via ducting. The overhead or underground service leads are taken into the Electricity Board's main fuse. This is usually 100 A for domestic purposes. From this fuse point the supply is taken to the meter and when three-phase supplies are provided, three main fuses are used with one composite meter.

With underground services, either the lead sheath of the cable or, as in the case of plastic cables, the wire armour is used to provide an earth. A separate wire is generally bound and soldered to the sheath or armour at the terminal position and then taken to an earth connector block. All protective conductors in the property are taken back to this block.

With overhead services, an earth block may be provided if protective multiple earthing (PME) is adopted. In other cases an earth electrode, in conjunction with a residual current circuit breaker, must be used. In some cases a separate overhead earth conductor is provided.

In large blocks of flats or offices the services to each floor are provided by 'rising mains'. These will not normally be encountered in domestic work. In cases where anyone involved with security systems finds rising mains it will be seen that the consumers' meters are located on the individual floors and the rising mains are used to carry the bulk supply up the building. Subservices will be teed off at the various floor levels.

Consumers' meters will be arranged to avoid the need to enter the premises. They may be visible through a small vandal-proof window or in a communal area.

It follows that in the domestic sector in the vast majority of cases only a single-phase 240 V supply will exist but in other applications multiphase supplies will be found. To this end it is vital that the supply is inspected and tested as previously described. Extra precautions must be taken to ensure that work is only being performed across one phase. Equally, when remote power supplies or separate mains feeds are being taken to different points of a system, it is advisable that they are all on the same phase. For example, with CCTV cameras that are mounted at various locations in a premises and are mains powered, they are all best powered from the same phase. This is more easily achieved if one circuit is run between the units with spurs installed local to each camera.

The protection of circuits and apparatus at the source can be done in a variety of ways, as outlined below.

Fuses

These are in line with the phase conductor and operate by opening the circuit when the fuse element melts following the heating effect of excess current. They may be semi-enclosed (rewireable) or cartridge in type.

Semi-enclosed

These have a fuse holder or link of an incombustible material such as porcelain or moulded resin. A fine wire runs between the contacts, partly enclosed in the holder or in an asbestos tube. The rating is governed by the gauge of the fuse wire link.

Cartridge

These have a similar fuse holder or link but the fuse element is contained in a cartridge of incombustible material filled with fine arc-suppressing sand or a similar material.

In practice the fuse will be installed in a fuseboard or consumer unit with the main isolating switch adjacent to the meter.

Miniature circuit breakers

Used as an alternative to fuses, these MCBs operate automatically and go open when the current flowing through them exceeds their preset value. The various ratings are determined by the mechanism involved which is usually based on an electromagnet and bimetal strip. They are installed in a distribution board in a similar manner to fuses; indeed some boards will accommodate the use of either.

In the electrical system the supply ends of the various circuits are brought back to a convenient point in the building and are then connected to a distribution fuseboard. Each individual circuit is protected by either a fuse or MCB. In the domestic environment there will normally be only one distribution fuseboard or consumer unit. The supply is controlled by a consumer's main double pole switch for isolation purposes. A pair of mains cables sufficiently large to carry the maximum current of the installation is connected from the Electricity Board's service fuse and neutral link through their meter to the consumer's main switch.

In a larger installation the security installer may find several distribution fuseboards, each supplying one floor or section of the premises. In such cases the incoming mains are taken to a main distribution board, where they connect to a number of large fuses or circuit breakers protecting the outgoing circuits. Submains cables connect the main board to a smaller branch distribution board or boards which will contain smaller fuses or MCBs that protect the actual final circuits.

The security system will therefore be derived at the consumer unit in one of a number of ways; however, at the extreme end it is most likely to be only connected by a spur. This is available as a metal-clad or insulated device, flush or surface mounted. All types can be used with a standard pattress. It is important to ensure that the outgoing side is wired to the out or load terminals, with the supply from the consumer unit connected to the inside.

6.2 Power supplies

These will be found powered by control equipment or will be remotely located. Most often they also have a battery standby to support the system in the event of the mains failing. If they are to be used in conjunction with remote signalling they must be type approved by the telephone authority.

The term 'power supplies' can mean a transformer, a battery or a rectifier filter, with or without a charging circuit that converts ac to dc, but we tend to apply the term to the components as a group. In most instances the power supply will be found to incorporate rechargeable batteries of sealed lead acid (SLA) form.

Figure 6.1 illustrates the power supply architecture. The power supply starts at the step-down transformer which converts its single-phase mains supply of 240 V ac to, typically, 12–24 V ac. The transformer is a device that uses electromagnetic induction to transfer electrical energy from one circuit to another, without direct connection between them. In its simplest form it consists of separate primary and secondary windings wound on a common core of ferromagnetic material such as iron. When an alternating current flows through the primary, the resulting magnetic flux in the core induces an alternating voltage across the secondary, hence causing a current to flow in an external circuit. In the case of a step-down transformer, such as the type we are dealing with here, the secondary site will have fewer windings than the primary. From the transformer, power is provided through a two-conductor cable to a rectifier and filter where ac is converted to dc. A charging circuit within the power

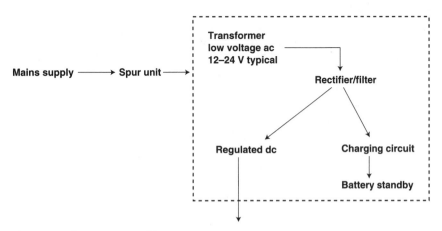

Figure 6.1 *Power supply architecture*

supply ensures that a standby battery may be constantly charged so long as ac is present.

The power supply must always be voltage regulated and be able to hold a fixed voltage output over a range of loads and charging currents. Microprocessor components, especially integrated circuits, are designed to operate at specific voltages and are not tolerant to fluctuations. Low voltages cause components to attempt to draw excess power, further lowering their tolerances, whilst higher voltages will actually destroy some components. For this reason the voltage should be measured at source and once again at the input terminals on the equipment.

A critical factor in the selection of a power supply is the determination of the load it must support. The first step is to establish how much power will be required by all power-consuming devices connected to the supply. It must then be calculated how long the standby supply must be able to satisfy the system if the primary supply is lost. We should state that by the primary supply we mean the electricity supply to the building which for the vast majority of the working time will support the system. The secondary supply is the batteries which provide the support if the mains supply is lost.

Essentially the mains supply must be a source that:

● will not be readily disconnected;
● is not isolated at any particular periods of time;
● is wired via a fused spur;
● is free from voltage spikes and current surges;
● is run to the equipment in a permanent fashion and not by a plug and socket or remote spur that can be switched off innocently or deliberately.

Of course the power supply may well be an integral part of the control equipment such as the intruder panel, fire panel or access controller, but in other instances it will be a stand-alone unit. In such a case it can be used to support auxiliary equipment such as additional powered detectors and remote signalling equipment, or to boost the branch voltage on multiplexed or addressable systems. Using electro-magnetic relays, the control panel trigger can switch the relay coil so that the power supply can generate extra potential to drive heavy duty devices.

The uninterruptible power supply may also be encountered. This has a greater ability to negate interference and mains supply surges, but tends to be used more in computer supplies that have back-up systems. It is often identified by the term 'monitored UPS'.

The reader should recall that the term power supply may well also be used to describe what is really only a small step-down transformer used to drive small intercom and access systems. In these systems the device will be a small insulated transformer providing low voltage ac for the speech circuits and for the door release. In a larger system or to drive heavier releases, a traditional power supply would be offered. The purpose of the small transformer type is to enable safe voltages to be adopted in the simple system without the requirement for a dc working with a standby battery

6.3 Batteries. Secondary supplies

Batteries are classified into two broad categories – primary (non rechargeable) and secondary (rechargeable). In the context of standby devices for power supplies, we are interested in sealed lead acid units which are also often called valve regulated devices. At other stages of the book we have come upon rechargeable batteries such as the Ni–Cad but it is the SLA battery that is the essential standby device in power supplies.

The sealed lead acid battery comprises positive and negative lead oxide material with a microporous glass fibre separator and an electrolyte of dilute sulphuric acid. The electrolyte is held immobile in the absorbent separator material. The first SLA batteries used a gel additive to immobilize the electrolyte, but as technology developed non-gel batteries became predominant and today this is the adopted method of construction.

Alternate oxidation and reduction of the respective lead components is achieved by releasing the potential as current in discharge and restoring it by applying current during charging. The battery is classified as secondary because energy must be applied before it can be used. Batteries are built up on a 2 V cell basis to give 2–4–6 and 12 V blocks. Built into each cell is the resealing vent which allows gas to escape if there is any overcharge, but which prevents electrolyte escaping under normal conditions. With the vent correctly in place a vacuum is created in the cell and a slight concavity may be seen in the walls of the battery. It is not possible to add water or electrolyte to a closed battery and it is vital to retain water in the system. The gases generated during overcharge are recombined so that they are not lost. Should oxygen and hydrogen escape from the battery a gradual drying out would occur, eventually cutting capacity and shortening life.

Batteries are rated in terms of their voltage and ampere hour capacity. The open-circuit voltage of any fully charged lead acid cell is just greater than 2.1 V so a 6 V battery with three cells in series would have an open-circuit voltage of 6.3 V and a 12 V six cell battery would be 12.6 V.

Capacity in terms of the battery is more complex. It is expressed in amp hours (AH) and is the total amount of electrical energy available from a fully charged cell. The actual amount of available energy is dependent on the discharge current, the temperature, the end or cut-off voltage and the general history and condition of the battery. For security installations SLA batteries are rated on the basis of a 20 hour constant current discharge at 20°C to a cut-off voltage of 1.72 V per cell. As an example, a 12 V 6.5 AH battery can discharge 325 mA (1/20 of 6.5 A) for 20 hours before the voltage drops to 10.32 V (6 × 1.72). The same battery will not, however, deliver 6.5 AH for 1 hour but will last about 30 minutes. When a battery discharges at a constant rate, its capacity changes according to the ampere load: capacity increases when the discharge current is less than the 20 hour rate and decreases when the current is higher.

For SLA batteries constant charging is needed. It is necessary to apply a voltage at the terminals greater than the voltage of the discharged battery. In a discharged state the battery voltage is lowered and is therefore most receptive to accepting current. If the charger output is great enough the battery will receive current faster than its capacity to use it in the recharge chemical reaction. The result is that the excess charge current is converted to heat which, if applied for too long will damage the battery. In cases where a charging circuit can put more than 25–30 per cent of the rated capacity into the battery then some form of current limiting is desirable. As an example, with a 6.5 AH battery the limiting current should be no more than 30 per cent of 6.5 A or 1.95 A. For float use in the security industry, the peak voltage output on the charging circuit should not exceed 2.25–2.3 V per cell or 13.8 V for a 12 V six cell battery, which is the time-honoured standard.

It is only possible to devise battery chemistry systems where electrical leakage is kept to a minimum – it cannot be totally eliminated. Batteries are active, not passive, so chemical reactions always take place which reduces the amount of available energy. In general, the rate of self-discharge is in the order of 3.5 per cent per month but this is accelerated as the storage temperature rises and decreases as it drops. For this reason it is essential that batteries are date coded and stored in a cool environment.

The battery must by necessity be able to support the system to maintain continuous operation for a period depending on the system function following mains failure. For intruder alarms this is 8 hours. Special times also exist for fire systems and emergency lighting and these are covered in the relevant chapters as appropriate to the system type. In all cases, on installation the batteries are to be marked with the date. If the voltage goes low a signal should be given and it should not be possible to set any system with this low voltage.

SLA batteries will be found to vary in capacity from 0.8 AH to 130 AH and in general should be changed after a period of 5 years.

There are certain maintenance considerations that we can summarize with regard to SLA batteries for use with security systems:

- Always check the float voltage (constant voltage applied to the battery) prior to installation.
- Give a period of charge and then conduct a short discharge test.
- Record the installation date and float voltage.
- Never fully discharge a battery. If taken to zero volts and left off or on load the internal battery resistance will elevate and the battery will no longer accept charge.
- Adequate ventilation should be provided for the battery.
- Avoid soldering to the terminals.
- Avoid use outside of −15 to +50°C.
- Provide free space between batteries. Allow 5–10 mm.
- If two or more battery groups are to be used in parallel they should be connected to the load through lengths of wires with the same loop resistance.
- Do not store in a discharged state or use alongside batteries of a different capacity or a different age or which have been subject to a different use.
- Do not install close to equipment which is capable of producing electrical sparks as the battery may generate ignitable gases.

6.4 Cable protection

Good practice is required to ensure that the system is neat and that mechanical protection is afforded to vulnerable wiring. Regulations also cover the segregation of cables of different categories and the susceptibility to electromagnetic interference.

Systems must comply with appropriate standards in relation to the environmental conditions to which they are liable to be subject at the protected premises. These conditions extend beyond mechanical damage to problems that could occur as a result of the weather, heat, dampness, corrosion, oil or electrical interference, or because of an adverse industrial atmosphere. The security installer will be involved with many different cable types and environments and so should be aware of the various protection and fixing methods, as outlined below.

Steel conduit

Using this method steel tubes are fixed to the walls and building structure and the cables are drawn into them at a later stage in time.

Cables in steel conduits are usually PVC insulated. Although the tubes can be fixed to the surface of any building they can also be attached to a building during the course of erection so that the tubes are later encased in plaster. It is not good practice to draw the cables into the tubes until the plaster has dried, otherwise they can trap moisture.

Conduit is available in diameters of 20 or 25 mm, either black enamelled or galvanized, or for aggressive environments it may be made of high grade 316 stainless steel. The tubes are themselves screwed together by the use of threaded sockets, tee pieces or elbows. When completed the entire system should be electrically continuous so that a current applied to any part of the assembly will be conducted to the main earthing terminal of the installation.

Steel-screwed conduit tubing gives a time-honoured and essentially sound system. There is little to remember regarding its use other than that it is important to avoid overcrowding of cables as excessive strain applied during the drawing process may stretch and fracture strands of the cables. This will effectively reduce the current-carrying capacity and the insulation may also be damaged, so that the cable is rendered useless. In well planned installations it should actually be possible to draw any cable from the conduit without damaging others.

Non-metallic conduit

Steel conduit tubing provides excellent protection for all cables and gives earth continuity, but can contribute to condensation and the corrosion of the material during the life of the installation. This can also lead to the loss of efficient electrical continuity. For the reasons given it is always worth considering non-metallic conduit tubing which is available in either rigid or non-rigid forms. Both consist of a PVC formulation, providing high impact resistance, and are of the same diameter as their steel counterparts. PVC conduits are now well established and have some particular advantages, including:

● High resistance to corrosion by water, acids, alkalis and oxidizing agents and the elements encountered in concrete and plaster.
● High dimensional stability and non-ageing.
● Incapable of supporting combustion.
● Not susceptible to water condensation.
● Has excellent electrical properties with an electrical breakdown voltage of the order of 12–20 kV/mm.

The rigid form is the most common and has plain bored ends with either heavy or light gauge walls and comes in black or a white self-coloured form. In practice it is installed on surface installations where mechanical damage is a distinct risk. The normal method of jointing and applying fittings is by the use of push action, in that the conduit is put into the receptive unit by a pushing force. This ensures a tight reliable fit, although adhesive can also be used with solid rubber gaskets for damp conditions. Threading is not employed as it weakens the walls.

The non-rigid type is found in long coiled lengths and is used for sunk or concealed wiring where appearance is not important. It has great flexibility to enable bends to be easily negotiated and it may also be threaded through holes. These types of tube are joined by using a small length of the conduit as a sleeve and pushing the lengths together using the sleeve as a restraint. The seal is then supplemented with a suitable compound. This form of tube is often found buried in floors before screeds are applied. Metallic and non-metallic conduits of rigid form can be used with a vast array of fittings and accessories, including couplings, reducers, bushes, tees, bends and elbows, and with a variety of junction boxes.

There are few disadvantages of PVC conduits but they must be considered with care if temperatures are below –5°C or above 70°C for extended periods of time. In low temperatures the PVC becomes hard and less ductile and can crack if subject to a hard blow. In higher temperatures the PVC can expand. For this reason expansion couplers should be used when rigid PVC tubes are installed in straight runs for lengths in excess of 5.5–6.5 m. An exception is where bends are employed as these tend to compensate for any expansion. The saddles that are used to support the conduit do themselves permit a measure of lateral movement.

Trunking

The erection of conduit tubing must be affected prior to running in the cables; however, there exists an alternative form of cable protection which allows cables to be installed before the protection. This is PVC trunking and it can be applied at the very last stage of an installation or may be applied to an existing non-protected cable system.

Trunking is a form of channel manufactured from high-impact PVC and having a locking or double-locking lid that is pushed into position and then held within the longitudinal channel ridges. Generally found in self-coloured white, it has a good aesthetic appearance and is ideal for most types of electronic protection and security systems. It is manufactured to BS 4678 and may be held in position by fixing its back-channel section in place with screws or it may be of a self-adhesive backed variety. Outlet boxes to complete the installation are available to ensure an adequate level of mechanical protection, as required by the IEE Wiring Regulations. There is a range of sizes available but the following are the most popular in common use: 16 mm (W) × 12.5 mm (D); 16 (W) × 16 mm (D); 20 mm (W) × 10 mm (D); 25 mm (W) × 12.5 mm (D); 25 mm (W) × 16 mm (D); 38 mm (W) × 16 mm(D); 38 mm (W) × 25 mm(D); 38 mm (W) × 38 mm(D).

As with conduit there is a vast range of accessories to make up the installation and blank ends can be adopted to terminate runs and to enclose the end of the trunking for both strength and aesthetic purposes.

Mini trunking can also be acquired which comes in a dispenser box. It is fitted to the required surface in a flat form and the sides may then be folded up and the lid clipped into place. Once again accessories to complete its installation are available.

Galvanized trunking is available but is mainly used in the industrial sector to protect mains cables. This version has the lid fastened by integral fixing bars that engage the trunking body when the captive lid screws are rotated through 90°. Another product encountered is the flange tray, carrying heavy cables. Signal cables are never to be run in close proximity to either galvanized trunking or flange trays when holding mains supplies, and cables carrying any security audio or video transmission should also follow a different route.

Aluminium tubing

This is a further form of cable protection using a tube of 12.5 mm diameter. It is in fact a long-established method for protecting taut wiring used as a detection device for openings such as windows. It may be found protecting individual cables in some installations where trunking is not practical, such as in external applications where the superior appearance of tubing and its total enclosing properties are desired. Aluminium tubing does not corrode and is suitable in harsh environments. It is joined by clamp couplings that feature screws and bolts that clamp the coupling ends over the tubes being joined. Elbows to negotiate bends and changes in direction are also applied by a clamping technique, with saddle clamps being used to fix the tubes to the building structure.

Channelling/capping

PVC-sheathed cables intended to be buried in the plaster of a building are better protected if first a PVC channel or capping is applied over them. This form of protection is generally used for the protection of electrical and security installation cables when surface wired to brickwork or blockwork prior to plastering or rendering. It is supplied in PVC material in a variety of lengths and sizes and is so flexible that it can be carried in reels and is shatterproof so does not crack when nailed in position. It readily takes up the contours of the wall and is cut to length using cable cutters or shears. It is insulated electrically and is self-extinguishing with low smoke, low toxicity and low emission properties. It is also a great aid to the plastering process in that the cables are held securely within its confines before plastering commences. In practice most capping used is of the PVC form but metal capping can also be found.

We can summarize by saying that we have now considered steel conduit systems, non-metallic (PVC) conduit systems, trunking, aluminium tubing and channelling/capping, all of which have a role to play in the protection of cables. With some of these systems the protection is there to support the cables prior to plastering or rendering, but with others it is the final method of giving protection and making up a neat cabling network.

The principal cable insulation that we meet is PVC (polyvinyl chloride) which does of itself resist attack by most oils, solvents, acids and alkalis. Equally it is not affected by the action of direct sunlight and is non-flammable and so is suitable for a wide range of internal and external duties. PVC-insulated cables can be run between floors and ceilings, and holes made for the passage of cables through ceilings can be made good by filling with cement or another building material as a precaution against the spread of fire. However, there is still a need for all cables involved in security roles to be well protected, be it the mains cable for safety or the signal cable for security.

For all systems the entire cable network must be protected from all likely damage, including mechanical, electrical and environmental. The interconnecting cables must be adequately supported and their installation must conform to good working practice. Interconnecting wiring must not be run in the same conduit or trunking as mains cables unless they are physically separated. The IEE Wiring Regulations do not permit the running of extra low voltage cables with mains cables unless the insulation resistance of both are equal; however, it is usual practice not to run signal cables alongside mains cables or to run them through the same holes in building structures or to feed them through the same entry to a control panel or power supply.

In some instances wiring must be screened and protected from radio and electrical interference. The problem in the industrial and commercial sectors with regard to interference are always much greater, but the demand for cable segregation applies to all applications.

Within this book we are dealing with a wide variety of security system types and the demands for cable protection do differ somewhat. However, we can agree that for all systems protection must be afforded. In some areas the need to protect from mechanical damage is not essential since cables may not be accessible or they may be concealed, but they still need to be supported. Table 6.1 can be used as a guide for the spacing of cable supports for PVC cables when they are actually in accessible positions but there is no practical need for further protection or enclosing of the cables. These spacings may be used to give good fixing positions, although they may be modified slightly by certain Regulatory Bodies if they feel other considerations are to be taken into account.

Table 6.1   *Spacing of cable supports for PVC cables in accessible positions*

Overall cable diameter (mm)	Horizontal (mm)	Vertical (mm)
<9	250	400
>9, <15	300	400
>15, <20	350	450
>20, <40	400	550

Table 6.1 refers to cables supported by clips or saddles. PVC clips with a single hole fixing are the most popular, with the internal surface of the clip being formed to suit the cable outside size and form. Self-adhesive clips may also be used where the surface being fixed to cannot easily be adapted to suit clips or screws.

In the event that cables are installed in areas not normally accessible and they are resting on a reasonably smooth surface, then no fixing is needed. However, on vertical runs fixing needs to be effected on lengths in excess of 5 m.

When running in cables the holes are to be made good and tubing applied when negotiating sharp surfaces including passing over the edges of bricks. Cables under floors must not be installed so that they can be damaged by contact with the ceiling or floor or their fixings. Cables passed through drilled holes in joists must be at least 50 mm vertically from the top or bottom and be supported by battens over extended runs. When passed through structural steelwork the holes must be fitted with bushes to prevent abrasion.

Cables should also follow contours, travelling in straight lines and never diagonally across walls. They are not to be allowed to pass close to steam or hot water pipes and must be at a sufficient distance to prevent any rise in surface temperature above the cable jacket's designed ambient temperature.

The final point to consider with respect to cable protection is that pertaining to safe earthing and screened cable.

6.5 Earthing and screening

At the beginning of the chapter we found it prudent to consider earth continuity and earth loop impedance, since the size of the earth conductor and properties of the earth must be adequate to carry sufficient current to blow a fuse or operate a cut-out device in the event of a fault.

Safety of the mains supply is achieved by ensuring that the mechanical protection of the cables is adequate and that the insulation between conductors and between the conductors and earth is within the requirements of all regulations. We can also reiterate that the continuity of the earth conductor must not of itself present any resistance in excess of a given value.

Earthing conductors are, in effect, a bond between all exposed metal in a premises and the final earth position which ensures that an electrical path exists that will operate current leakage protection in the event of a short circuit. The removal of an earth conductor can mean that part of a system can become live. However, the reference earth is different as it is only a path for induced current to flow from, for example, the braid of a shielded cable. Such reference points are deliberately taken to earth at one point to ensure that earth loops are not set up in the shield. However, these are small in size compared to the main earth conductor needed to carry sufficient current to blow a fuse or operate a residual cut-out device. This can be seen by reference to Figure 6.2.

We find that in semiconductor electrics low values of current exist and, because short lengths of leads and PCB tracks are involved, only negligible voltage drops occur. However, when we consider interference signals, the current and hence the voltage drop in earth leads can be much higher than for normal equipment working levels. To avoid any malfunction from interfering signals it is a help if all the leads to ground are taken to a single point, including the power

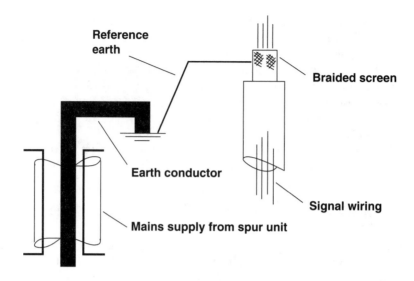

Figure 6.2   *Relationship of reference earth to main earth conductor*

supply. For this reason the earthing is generally at the power supply and any screening used to protect cables from interference is also connected to that point. To maintain single-point earthing the screen has to be sheathed in an insulating sleeve to stop the screen contacting a further earth medium and forming an earth loop.

All those involved with security systems are becoming increasingly connected with remote signalling and data transmission due to the integration of methods. To this end a greater use will be found for screened cable, extending to data communications and the need to negate interfering signals.

In communication practices cables and individual conductors may be wrapped with a variety of materials. Such wrappings and jackets include insulation and shielding and strengthening that protect the assembly against stretching during installation. The materials are there to combat dampness, protect from cold and heat sources and to provide shielding from electromagnetic interference. In communications the conductor cables contain copper wires and require a ground wire or grounding material within the jacket in addition to the wires that carry the current. The exception is fibre optics which do not need a ground.

Fibre optics have already been considered in their own right, so we shall here overview the other three principal forms of cable.

Unshielded twisted pair (UTP)

Here the cable holds a number of pairs of twisted wire. Each pair is twisted differently and is defined by the number of twists over a given length. The wires are not run parallel with each other inside of the outer jacket but twist across each other as this helps to cancel noise or electrical signal interference which may be induced by adjacent cabling feeding different apparatus.

Shielded twisted pair (STP)

Once again the jacket holds a number of pairs of twisted wires but these are themselves wrapped within a foil inner jacket. The sets of foil jackets are wrapped together inside a braided copper mesh and this is all enclosed within the outer jacket. This assembly form is generally used in computer networks.

Coaxial cable

Common in CCTV and observation systems, this can also be found in computer and data transmission networks. It has a copper core which may be single or stranded, with the latter type employed in applications which may involve greater cable bending. The second conductor is the shield which surrounds the core. The core is packed in plastic insulation which is then wrapped in the shield of braided copper mesh. The assembly is then contained within an outer jacket. The connector will be found to be essentially tube-like and to have only one pin.

We can therefore draw the conclusion that the needs for electronic protection and security systems will forever remain, but they will increasingly appear in diverse forms ranging from the small device to the large complex systems involving integration. Within these systems there will be a huge variety of cable types, so that the wiring will extend beyond security to safety also, with mains and earthing connections.

In real terms the subject remains of vital importance to us all, ranging from the manufacturer to the installer and end user. The responsibilities considered apply equally.

7 Reference information

7.1 Audible signalling devices

The first section of reference information is dedicated to audible signalling or warning devices because of their widespread use throughout the electronic protection field. Many items of security apparatus have integrated audible devices to draw an operator's or user's attention to a signal having been generated, so it is only sounders in common use and which can be employed across the spectrum that we wish to consider. In all cases these are available to work at different voltages and at various sound levels. There are other facts that the reader should be aware of when selecting audible signalling devices, as outlined below.

Electromagnetic bell

Here the current is applied to a coil or solenoid and the plunger is attracted through its centre to strike a dome. As it strikes, a pair of contacts go open circuit so that the current is interrupted and the plunger is returned to its free rest position by a spring. The contacts reclose and the process is repeated. The distance travelled by the plunger determines the 'ring' and is adjusted by a screw or the rotating of an eccentric dome.

Centrifugal bell (motorized)

This device uses a small motor to rotate a striker at high speed which repeatedly strikes the inside of the bell dome. The striker's direction and movement is influenced by centrifugal force. The current drawn by the motorized bell is generally marginally higher than that of the electromagnetic bell. Once again the level of ring can be adjusted.

Solid state electronic horns

These comprise an oscillator, amplifier and horn loudspeaker, all contained within the same unit. The horn is re-entrant, in that it is folded within itself to save space. These are highly directional devices and need to be aimed at the area in which they must be effective. The frequency of pitch will vary between models. The horn is popular as it is inexpensive, delivering 120 dB(A) and drawing only some 400 mA at 12 Vdc.

Piezo siren

This is a high output siren with a high pitch and low current consumption. The principal element is a ceramic piezo transducer which generates an audible tone and frequency when energized by a peak square-wave driven by a low power consumption IC. The transducer is mounted rigidly in its housing. Often found as an internal sounder.

Mylar cone speaker

Often found where multiple tones are required, these speakers may be mounted on a square or round chassis and range in size with outputs up to 100 dB(A) at 1 m. The power output can be from 0.5 to 15 W. The system comprises a mylar cone speaker which is a transparent film unaffected by moisture. Ferrite magnets with a plastic centre dome combined with a card gasket make the speakers weatherproof from the front. Capable of good low frequency, long range transmission, they are still classed as being of solid state construction even though they have a speaker operation.

Mechanical siren

This consists of an electric motor which drives an impeller that in turn forces air through vents in the casing in such a way as to create a loud sound. Semi-enclosed, they are for use in external applications but not indoors as the impeller should not be accessible under normal circumstances. Their sound output is high but the running current is much greater than that of solid state electronic sounders and the in-rush current considerably more. They are often found as mains sounders powered via a relay

Selection considerations:

- SAB or SCB modules integrated with the sounder provide monitoring and battery standby support.
- Use steel shroud covers/enclosures over the sounders if blowtorch attacks could be carried out.
- Twin steel shroud enclosures provide drill protection by electrically detecting the drill bit on piercing and shorting the covers together.

- Louvreless covers prevent the injection of sound-deadening foam. Polycarbonate covers provide reasonable protection from impact and abuse.
- Tamper protection should prevent front screw or rear tamper and unit removal.
- High pitch tones are excellent for short range duty but do not carry well over long distances.
- Low pitch tones provide good long range transmission. Frequencies in the order of 800 Hz to 1 kHz carry effectively at extended distances.
- For better identification of an alarm sounder, the assembly is best complemented with a visual signalling xenon flasher/ beacon.
- For high security risks use steel louvreless double-skinned sounders with an SCB module and a pull-up resistor to produce an alarm if the trigger wire from the control equipment to the sounder is cut. Include full tamper protection including back tamper to stop the unit being removed from its fixing.

Propagation of sound:

- 100 dB(A) is considered loud.
- 130 dB(A) is the threshold of pain and will cause permanent hearing damage.
- Loudness does not follow a linear scale – a sound measuring twice the level of another does not appear twice as loud to the observer. Progression is in a logarithmic manner.
- Installing two or an even greater number of sounders in close proximity to each other and of the same type will result in only a marginal increase in audibility.
- Sound levels decrease with distance from source, generally at a rate of 6dB(A) for a doubling of the distance from a point source.
- Sound is affected by any breeze. The range is increased in the direction to which it is blowing and decreased in others.
- The normal temperature gradient, whereby air gets cooler with height, causes sound to refract upwards and be reduced at ground level, thus lowering the range.
- In cases of temperature inversion, when the air above the ground is warmer than at a lower level, the sound waves are bent downwards, increasing the range.
- Sound levels decrease with distance and most sounders are specified at 1 m. Sound pressure for most sources drops in proportion to distance. At twice the distance the pressure is reduced to a half and at three times it is a third.

● A sounder at a height of 6 m is only one-third of its quoted rating at 1 m (approximately 16 dB(A) less).
● The minimum sound level at which a sounder can be deemed audible above moderate ambient noise is 60 dB(A).

Installation of external sounders:

● Select an area as inaccessible as possible.
● Fix to a solid monolithic structure of concrete, brick or hardwood.
● Power will be lost if the sounder is capable of vibrating the building structure.
● The direct area is best free from high, dense hedges or trees as these will attenuate the sound. Sound reduces by 6 dB(A) for every doubling of initial measurement distance on flat open sites, and by up to 10–12 dB(A) more in areas of vegetation.
● If fixing the sounder to any panel install a back plate and use bolts with the heads covered by the sounder enclosure.

7.2 Electromagnetic relay selection

Electromagnetic relays can be used to interface systems or switch greater loads through the main contacts. There are four particular classifications:

● *General purpose relays*: Here the contacts turn on instantaneously when the coil is energized and go off when de-energized.
● *Plastic sealed relays*. The mechanism is encapsulated in a plastic case with the terminals and terminal block sealed by epoxy resin.
● *Hermetically sealed relays*. The internal mechanisms are completely sealed from the external atmosphere by a metal case and metal terminal block.
● *Power relays*. These are intended to switch heavy loads.

The second classification has two further categories:

● *Latching relays*. The contacts of these relays are magnetically or mechanically locked in either the energized or de-energized position until a reset signal is applied.
● *Ratchet/stepping relays*. The contacts of these relays alternately turn on and off or sequentially operate when a pulse signal is applied.

Having selected the classification, the next stage is to determine the voltage to operate the coil. It may be ac or dc. Table 7.1 gives some reference values.

Table 7.1   *Electromagnetic relay coil voltages*

Coil voltage	ac	6 V	12 V	24 V	–	50 V	220/240 V
	dc	6 V	12 V	24 V	48 V	–	220/240 V

Next, one determines the contact materials, a decision which is based on the load current and resistance to corrosion and depending on the area the system must operate within. Table 7.2 gives some examples.

7.3 Resistors

Resistors are designed to introduce a specific amount of resistance into a circuit. They are available in many physical sizes and with a range of ratings, and can be grouped into two main types:

- *Fixed resistors.* These have values that do not change and the actual value remains relatively constant unless the device is damaged. They exhibit some resistance changes in response to temperature fluctuations but these are largely insignificant and may be disregarded.
- *Variable resistors.* These may be called rheostats or potentiometers and are adopted when it is necessary to change the resistance in a circuit.

Combining resistors

It is possible to increase or decrease resistance by using a number of resistors in conjunction with each other. In series the sum is additive, i.e. $R_t = R_1 + R_2$. Hence 1000 Ω can be obtained by using two 500 Ω resistors in series as the current must flow through both. In parallel the sum is

$$R_t = \frac{(R_1 \times R_2)}{(R_1 + R_2)}$$

for two resistors and:

$$R_t = \frac{1}{(1/R_1) + (1/R_2) \ldots + (1/R_n)}$$

when several resistors are connected in parallel, where R_n is the total number of resistors.

7.4 Diodes

Diodes only allow current to flow in one direction. Semiconductor diodes break down at certain voltage levels and the specification will list the peak inverse voltage (PIV) or peak reverse voltage (PRV). Two types will be mentioned here:

- *Zener diodes.* These differ from regular diodes and respond to reverse voltages in a special fashion. The voltage rating of a zener diode is the point at which it begins to conduct when reverse biased.
- *Light emitting diodes* (LEDs). These conduct in one direction, emitting light when forward biased. Common colours are red, green, yellow/amber and blue.

They are a durable component intended for low voltage circuits and most operate on 3–6 V dc. Since they are forward biased they can be used to check polarity or operate an indicator as shown in Figure 7.1.

7.5 Capacitors

Capacitors store voltage until a threshold is reached when the voltage is released. The unit of capacitance is the farad but the microfarad, μF, is used to rate the capacitor. They have two voltage levels, the

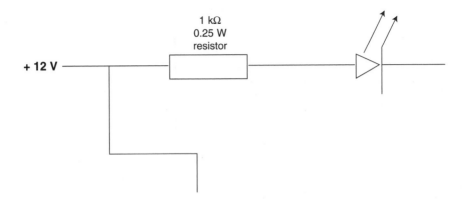

Figure 7.1 *LED indicator on a 12 V dc circuit*

Table 7.2   *Electromagnetic relay contact materials.*

Low load current ◀─────────────────────────────────────▶ High load current

PGS alloy (platinum, gold silver)	AgPd (silver palladium)	Ag (silver)	AgCdO (silver cadmium oxide)	AgNI (silver nickel)	AgSnin (silver, tin, indium)	AgW (silver tungsten)
High resistance to corrosion. Mainly used in minute current circuit	High resistance to corrosion and sulphur	High conductance and thermal conductance of all metals. Low contact resistance	High conductance and low contact resistance like Ag with excellent resistance to metal deposition	Rival with Ag in terms of conductance. Excellent resistance to arc	Excellent resistance to metal deposition and wear	High hardness and melting point. Excellent resistance to arc, metal deposition and transfer, but high contact resistance and poor environmental durability

breakdown voltage and the working voltage. The breakdown voltage is the maximum that the dielectric insulating material can withstand without breaking down. The working voltage is the maximum voltage that can be safely placed across the capacitor plates or electrodes. There are several types of capacitors:

- *Ceramic capacitors*. These have a wafer of ceramic material between two silver plates. Working voltages range from 50 to 1600 V.
- *Mica capacitors*. These have strips of mica between interconnected plates and have values across the range from 5 pF to 0.01 μF and working voltages from 200 to 50 000 V.
- *Paper capacitors*. These have waxed paper between two strips of tin foil. Their range is of the order of 0.0001 μF to 1 μF with working voltages from 200 to 5000 V.
- *Electrolytic capacitors*. These are polarized for use in dc circuits and have aluminium foil plates. Working voltages are from 3 to 700 V or even greater. Their range is of the order of 0.47 μF to 10 000 μF.

7.6 Multiplication factors

When dealing with electrical circuits the values of current, voltage and power are often expressed in multiples of their units by addition of a prefix letter. Some commonly used multiples are given in Table 7.3, while Table 7.4 shows the range of multiplication factors.

Table 7.3 *Common multiples*

Unit	Multiple	Value
Ampere	Milliampere (mA)	1/1000 ampere
Ampere	Microampere (μA)	1/1 000 000 ampere
Volt	Millivolt (mV)	1/1000 volt
Volt	Microvolt (μV)	1/1000 000 volt
Ohm	Kilohm (kΩ)	1000 ohms
Ohm	Megohm (MΩ)	1000 000 ohms

Table 7.4   *Multiplication factors*

Multiplication factor	Prefix
1000 000 000 (10^9)	Giga (G)
1000 000 (10^6)	Mega (M)
1000 (10^3)	Kilo (k)
100 (10^2)	Hecto (h)
10 (10)	Deca (da)
0.1 (10^{-1})	Deci (d)
0.01 (10^{-2})	Centi (c)
0.001 (10^{-3})	Milli (m)
0.000 001 (10^{-6})	Micro (μ)
0.000 000 001 (10^{-9})	Nano (n)
0.000 000 000 001 (10^{-12})	Pico (p)

Index